The Urban Book Series

The Urban Book Series is a resource for urban studies and geography research worldwide. It provides a unique and innovative resource for the latest developments in the field, nurturing a comprehensive and encompassing publication venue for urban studies, urban geography, planning and regional development.

The series publishes peer-reviewed volumes related to urbanization, sustainability, urban environments, sustainable urbanism, governance, globalization, urban and sustainable development, spatial and area studies, urban management, transport systems, urban infrastructure, urban dynamics, green cities and urban landscapes. It also invites research which documents urbanization processes and urban dynamics on a national, regional and local level, welcoming case studies, as well as comparative and applied research.

The series will appeal to urbanists, geographers, planners, engineers, architects, policy makers, and to all of those interested in a wide-ranging overview of contemporary urban studies and innovations in the field. It accepts monographs, edited volumes and textbooks.

Indexed by Scopus.

More information about this series at http://www.springer.com/series/14773

Ellen Hoffmann · Andreas Buerkert ·
Stephan von Cramon-Taubadel ·
Kotrakere Basegowda Umesh ·
Prasannakumar Pethandlahalli Shivaraj ·
Prem José Vazhacharickal
Editors

The Rural-Urban Interface

An Interdisciplinary Research Approach to Urbanisation Processes Around the Indian Megacity Bengaluru

Springer

Editors
Ellen Hoffmann
Organic Plant Production
and Agroecosystems Research
in the Tropics and Subtropics
University of Kassel
Witzenhausen, Germany

Andreas Buerkert
Organic Plant Production
and Agroecosystems Research
in the Tropics and Subtropics
University of Kassel
Witzenhausen, Germany

Stephan von Cramon-Taubadel
Agricultiral Policy
Agricultural Economics and Rural
Development
Georg-August-University Göttingen
Göttingen, Germany

Kotrakere Basegowda Umesh
Agricultural Economics
University of Agricultural Sciences
Bangalore
Bengaluru, India

Prasannakumar Pethandlahalli Shivaraj
Agricultural Economics
University of Agricultural Sciences
Bangalore
Bengaluru, India

Prem José Vazhacharickal
Rural-Urban Center
University of Agricultural Sciences
Bangalore
Bengaluru, India

ISSN 2365-757X ISSN 2365-7588 (electronic)
The Urban Book Series
ISBN 978-3-030-79971-7 ISBN 978-3-030-79972-4 (eBook)
https://doi.org/10.1007/978-3-030-79972-4

This Springer imprint is published by the registered company Springer Nature Switzerland AG
The registered company address is: Gewerbestrasse 11, 6330 Cham, Switzerland

Preface

Comprising a consortium of more than 70 scientists in 23 projects, and co-funded by the German Science Foundation (DFG) and the Indian Department of Biotechnology (DBT), the Indo-German collaborative project FOR2432 is a long-term research initiative dedicated to the processes unfolding in the rural–urban interface under the pressure of rapid urbanisation. The first Indo-German Conference on Rural–Urban Transitions, hosted in Bengaluru, India, from 20 to 22 February 2019, concluded the first of potentially three research periods and presented a first overview of the effects of urbanisation on the ecological, economic, and social conditions in that space. The volume presented here combines selected conference papers with some invited, complementary contributions. Like the conference, this book reaches out to the wider research community interested in urbanisation and Asian megacities, aiming to initiate an inter- and transdisciplinary discourse on rural–urban development paths towards a sustainable future.

Witzenhausen, Germany | Ellen Hoffmann

Contents

Part I Introduction

1 **The Rural–Urban Interface: An Interdisciplinary Research Approach to Urbanisation Processes Around the Indian Megacity Bengaluru** 3
Ellen Hoffmann, Stephan von Cramon-Taubadel, Andreas Buerkert, K. B. Umesh, P. S. Prasannakumar, and Prem José Vazhacharickal

Part II Changing Patterns of Land Use and Agriculture

2 **Temporal Dynamics of Land Use Change and Spatial Urbanisation Indices in the Rural–Urban Interface of Bengaluru Over the Past Three Decades** 21
Sheetal Patil, Thanh Thi Nguyen, B. Dhanya, Seema Purushothaman, and Ellen Hoffmann

3 **Urbanisation Effects on Agricultural Production Systems Across the Rural–Urban Interface of Bengaluru** 35
C. L. Nandini, C. Kavya, G. N. Nagaraja, and P. S. Prasannakumar

4 **Changes in Land Use Pattern Vis-a-Vis Urbanisation and Credit Across the Rural–Urban Interface of Bengaluru** 45
M. S. Udaykumar, K. B. Umesh, and P. S. Srikantha Murthy

5 **Dairy Production in an Urbanising Environment: A Social-Ecological Approach to Understand the Case of Bengaluru** 53
Marion Reichenbach, Ana Pinto, Sven König, Raghavendra Bhatta, and Eva Schlecht

6 **Capital Formation Across the Rural–Urban Interface of Bengaluru North: An Economic Analysis** 61
H. V. Harish Kumar and B. V. Chinnappa Reddy

Part III Resource Use, Efficiencies and Tradeoffs in Crop and Livestock Production

7 **Crop Production Under Urbanisation: An Experimental Approach to Understand and Model Agricultural Intensification** 71
Andreas Buerkert, Ellen Hoffmann, Renuka Suddapuli Hewage, Sven Goenster-Jordan, Suman Kumar Sourav, Andrea Mock, Prem José Vazhacharickal, C. T. Subbarayappa, Mudalagiriyappa, D. C. Hanumanthappa, Stephan Peth, and Michael Wachendorf

8 **Effects of Soil Management Practices on Soil Physical Properties and Water Cycle in a Multifactorial Field Experiment** 85
Albara Almawazreh, Daniel Uteau, Andreas Buerkert, Ayappa Sathish, Mudalagiriyappa, D. C. Hanumanthappa, C. T. Subbarayappa, and Stephan Peth

9 **Biomass Estimation of Vegetables—Can Remote Sensing Be a Tool for It?** 95
Thomas Astor and Michael Wachendorf

10 **Dairy Farms in an Urbanising Environment: Tradeoffs Between Productivity and Animal Welfare** 103
Ana Pinto, Tong Yin, Marion Reichenbach, Pradeep Kumar Malik, Eva Schlecht, and Sven König

Part IV Food Systems as Outcomes and Drivers of Rural-Urban Transitions

11 **Crop Diversity, Dietary Diversity, and Nutritional Intake: Evidence on Inter-Linkages Across the Rural–Urban Interface of Bengaluru** 111
Veerabhadrappa Bellundagi, K. B. Umesh, B. C. Ashwini, and Stephan von Cramon-Taubadel

12 **Food Security and Factors Influencing Food and Calorie Intake Across the Rural–Urban Interface of Bengaluru** 119
B. C. Ashwini, K. B. Umesh, Veerabhadrappa Bellundagi, P. S. Prasannakumar, and Stephan von Cramon-Taubadel

13 **Household and Individual Nutrition in the Rural–Urban Interface of Bengaluru** 127
Anjali Purushotham

14 **Nutritional Status of Women Across Rural–Urban Interface of Bengaluru** 135
K. Geetha, D. Vijayalakshmi, Shilpa Yatnatti, Christoph Dittrich, and Deepa S. Patil

15 "Homemade Food is Always Healthy": The Ambivalence of a Traditionalisation of Diets in Bengaluru 143
Mirka Erler, Neda Yousefian, and Christoph Dittrich

Part V Biodiversity and Ecosystem Functions Under Rural-Urban Transitions

16 Tree Species Diversity in the Southern Transect Across the Rural–Urban Interface of Bengaluru 151
B. N. Divakara, C. U. Nikhitha, M. A. Mahmud, Nils Nölke, and V. P. Tewari

17 Biodiversity of Birds Along the Rural–Urban Interface of the Indian Megacity Bengaluru 163
Ingo Grass, Arne Wenzel, Varsha Raj, Suryanarayana Subramanya, Catrin Westphal, and Teja Tscharntke

18 Bee Diversity in the Rural–Urban Interface of Bengaluru and Scope for Pollinator-Integrated Urban Agriculture 171
Chethana V. Casiker, B. Jagadishakumara, G. M. Sunil, K. Chaithra, and M. Soubadra Devy

19 Dynamics of Field Margin Vegetation with Changing Agricultural Landscape Across a Rural–Urban Interface 183
Sunil Nautiyal, Mrinalini Goswami, Ellen Hoffmann, and Puneeth Shivakumar

Part VI Conclusion

20 Dynamics in the Rural–Urban Interface of an Indian Megacity—Synopsis and Outlook 199
Ellen Hoffmann, Stephan von Cramon-Taubadel, and Andreas Buerkert

Part I
Introduction

Chapter 1
The Rural–Urban Interface: An Interdisciplinary Research Approach to Urbanisation Processes Around the Indian Megacity Bengaluru

Ellen Hoffmann, Stephan von Cramon-Taubadel, Andreas Buerkert, K. B. Umesh, P. S. Prasannakumar, and Prem José Vazhacharickal

Abstract Urbanisation is a global trend and a major challenge to human societies as well as to the planetary environment and resource pools, and the research landscape related to these issues is accordingly large and diverse. The collaborative Indo-German research project FOR2432 tackles urbanisation-related transformations by an interdisciplinary approach focused on the South Indian metropolitan area of Bengaluru. This editorial sketches the overarching research design, to which most of the book chapters refer, and describes some baseline characteristics of the research space. It thus illustrates the common background for the specific topics and targets of the subsequent studies collated in this volume.

Keywords Spatio-temporal research design · Research transects · Scaling · Baseline · Household surveys · Cropping profile · GIS-based mapping

---

E. Hoffmann (✉) · A. Buerkert
Organic Plant Production and Agroecosystems Research in the Tropics and Subtropics, University of Kassel, 37213 Witzenhausen, Germany
e-mail: ellen.hoffmann@uni-kassel.de

A. Buerkert
e-mail: buerkert@uni-kassel.de

S. von Cramon-Taubadel
Agricultural Policy, Agricultural Economics and Rural Development, Georg-August-University Göttingen, 37073 Göttingen, Germany
e-mail: scramon@gwdg.de

K. B. Umesh · P. S. Prasannakumar
Agricultural Economics, University of Agricultural Sciences Bangalore, Bengaluru 560065, India
e-mail: umeshuas@gmail.com

P. S. Prasannakumar
e-mail: shiprasannamk415@gmail.com

P. J. Vazhacharickal
Rural-Urban Centre, University of Agricultural Sciences Bangalore, Bengaluru 560065, India
e-mail: premjosev@gmail.com

E. Hoffmann et al. (eds.), *The Rural-Urban Interface*, The Urban Book Series,
https://doi.org/10.1007/978-3-030-79972-4_1

1.1 Introduction

How do agricultural production systems and household structures change at different stages of urbanisation? How does urban expansion affect the ability of regional ecosystems to provide food and other ecosystem services? How do exchange processes between agroecosystems, producers and consumers, or different social groups change as urbanisation advances? How do social and ecological systems interact where rural and urban livelihoods, traditions, aspirations, and forms of land use clash?—These are the key questions guiding the research presented in this book. Answering them requires a wide range of expertise that our current academic system segregates into different disciplines. Bringing highly specialised scientists back together, joining their forces in a common research space, and integrating their results to achieve a better understanding of the complex processes of urbanisation are the overarching objective in FOR2432, the project from which most of the contributions in this book emerged.

This kind of research quickly touches on the question of 'what is urban?'. Several of the studies here show that apart from 'urban' and 'rural' characteristics, 'transitional' areas have unique features that clearly distinguish them from either end of the spectrum, but defining a boundary is still not possible, as it varies with the parameters under investigation.

Following a systems approach, the papers investigate biophysical, ecological, economic, and social subsystems, at different (nested) scales, such as experimental plots, farmer fields, and transects across the rural–urban interface, or households, villages, and regionally operating companies and institutions. The interactions and feedbacks between these subsystems and scales lead to complex adaptive behaviour with ambiguous causalities. Decisions taken at a critical turning or branching point determine subsequent path dependencies. While contributing to problem solving and policy or management recommendations in some cases, the papers presented here aim at advances of basic knowledge, starting from a coherent description of the common research space.

Upon completion of a first wave of data collection, this book highlights major insights to rural–urban transition processes in Bengaluru that were gathered so far. These comprise intensification of agriculture or shifts from agricultural to other (urban) income sources, along with a diversification of livelihoods, food consumption patterns, and nutrition. The rural–urban interface features agroecosystems that harbour considerable biodiversity but are increasingly stressed by resource constraints, diverging attitudes and consumption behaviour among different social groups, and immense heterogeneity in spatial patterns of environmental and economic variables that tend to override a presumed uni-directional rural–urban gradient.

1.2 Bengaluru as an In Situ Laboratory for Research on Rural–Urban Transitions

The interdisciplinary research of FOR2432 was guided by the research questions outlined above, and by hypotheses derived from them. Its main operational pillars were the focus on a common research space, the spatially explicit of data collection, and a nested sampling design that allowed analyses at different scales. This formed the common background on which Bengaluru was considered an in situ laboratory of rural–urban transformation.

Bengaluru in the state of Karnataka, with 12.3 Mio people in 2020 (World Population Prospects 2019), is among the fastest-growing cities in India, and its metropolitan area currently covers 8005 km^2, comprising the districts of Bengaluru Urban and parts of Bengaluru Rural and Ramanagara (Fig. 1.1). The research space consists of two transects that cut across the rural–urban interface, one in the north (N-transect) and one in the south-west of the city (S-transect; Hoffmann et al. 2017).

The transects harbour 93 and 98 villages, respectively, of which 31 and 30 were selected as a representative sample for a comprehensive household survey, targeting a total of more than 1200 respondents. The sampling strategy employed an index composed of the proportion of built-up area and distance to the city centre as a proxy of the 'degree of urbanisation' for a given settlement (Survey Stratification Index, SSI; Hoffmann et al. 2017). The full range of SSI-values was subdivided into six

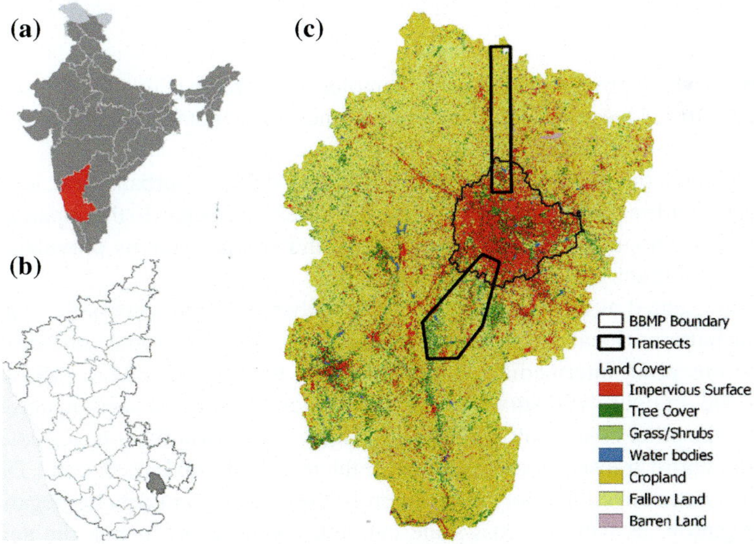

Fig. 1.1 Research space of FOR2432: **a** India with federal state boundaries, **b** Karnataka state (red area in **a**) with district boundaries, **c** land cover map of the districts Bengaluru Urban (grey area in **b**), Bengaluru Rural, and Ramanagara with boundaries of the Bengaluru Metropolitan Area (BBMP) and the northern and southern research transects (Map by Nils Nölke)

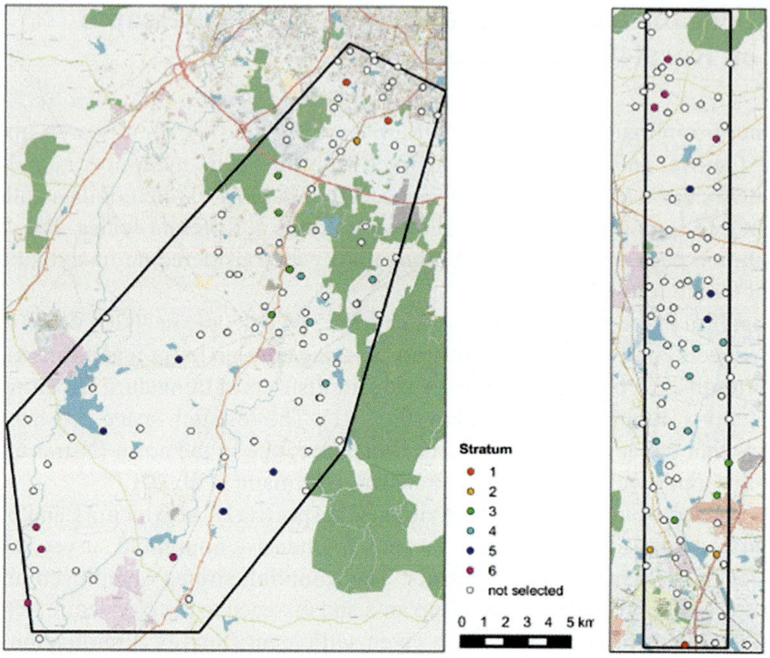

Fig. 1.2 Research transects and villages identified within them. Coloured dots indicate the random sample drawn for the household survey, with the respective stratum allocation (Hoffmann et al. 2017)

strata from which proportional random samples were drawn (Fig. 1.2). Subsets of the sampled villages were selected for various other, more detailed analyses presented in the subsequent chapters.

The SSI and the strata also served to tentatively delineate urban, rural, and transitional areas within the transects, or to correlate other parameters to the apparent rural–urban transition stage. Geographic locations and spatial patterns provided further means of alignment and integration of results.

Data presented at the village level, such as average socio-economic features or cropping patterns (local preferences or diversity of cultivated crops), provide discrete observations at an intermediate analytical scale. Primary data were collected at a lower scale, in household surveys, individual interviews, or on specific agricultural plots. On the other hand, remote sensing products such as satellite images potentially cover the entire transect areas and thus refer to a high analytical scale. Different means of data collection apply at different scaling levels, and data aggregation and disaggregation, as well as cross-scale calibrations are tools to link the analytical scales (Fig. 1.3). Usually, higher scale analyses combine with lower resolution data, and vice versa. Upscaling and downscaling, beside the interdisciplinary integration, are aims that recur throughout the chapters.

Analytical scale	Spatial dimension	Means of data collection	Data resolution	Image resolution (RS pixel size)
high			*low*	*low*
Rural-urban interface (Bengaluru Urban District)	2182 km²	Remote sensing (RS), Land cover analysis		
Transect	50 km²	RS, Land cover analysis		
Area, gradient section (rural / transitional / urban)		Geographic and/or statistical classification	Aggregation; Cross-scale calibration; Disaggregation	
Settlement (village, neighbourhood, ward)	~ 1 km²	Bhuvan map, official records of village authorities (panchayat, aganvadi), GIS mapping, focus group discussion		
Household / farm	~ 2 ha	Household survey		30 m (Landsat)
Plot / individual	6m x 6m	Household survey, Agronomic experiments, Biodiversity surveys		10 m (Sentinel)
Sample (soil / plant)	< 1 m²	Soil sensor records, Chemical analyses		0,3 m (Worldview 3)
low			*high*	*high*

Fig. 1.3 Levels of scaling and resolution in data acquisition and analysis of the rural–urban interface of Bengaluru, India

1.3 Baseline Characteristics Within the Research Transects

The large household survey conducted in 2017 provides the common background for the analyses presented in the subsequent chapters. This section summarises some descriptive survey results to illustrate the overall conditions in the research space and help to understand the conclusions drawn in the different contributions.

1.3.1 Social Strata in the Survey Sample

The household survey was designed to be representative of the society dwelling in the rural–urban interface of Bengaluru. The respondents covered the full range of social strata, and results mirror variations between the two transects, between rural, transitional, or urban areas, and between individual villages.

The assessment of the socio-economic strata of the household was based on the MRSI classification (The Market Research Society of India 2011; see also Chaps. 13–15). This scheme asks for the possession of different durable household assets (such as refrigerators or two wheelers) from a pre-defined list, as well as for the highest education of the household head. The scores are then combined in a matrix of 12 categories labelled A1-3, B/C/D1-2, and E1-3, of which the B and C categories are considered 'middle class'. Assets and education get slightly different weighting in this classification scheme. According to this broad definition, 92% of the surveyed households would categorise as middle class. To get a more differentiated and continuous picture, the sum of the asset score and the education score (social score) was applied here as a proxy to the original MRSI score, which reflects the distribution of social status in a given population.

In the entire sample of 1230 surveyed households, the social score ranged from 2 to 16 with an average of 8.47 (Fig. 1.4). The difference between the transects was minor, with a slightly higher share of lower scores in the S-transect (Fig. 1.4).

At the individual village level, the socio-economic profile looked quite distinct, as illustrated by three exemplary villages in the S-transect (Fig. 1.5). Muninagara, a rural (stratum 5) village where 22 households were interviewed, represented the lower end of the spectrum with an average social score of 6.14. Bettarayanadoddi, a village with transitional characteristics (stratum 4) and 20 surveyed households, had a score of 8.45 and was thus close to overall average (8.47). The urban neighbourhood Doddakalsandra (stratum 1) with 15 respondents scored on average 9.33 and thus represented the upper end of the range.

If the average social scores calculated for all villages were clustered around the full score numbers (i $\pm$ 0.5), and the villages in each cluster were counted, the distribution resembled that at the household level. Although the average of the social score was not clearly correlated to the SSI, a trend became apparent when the data

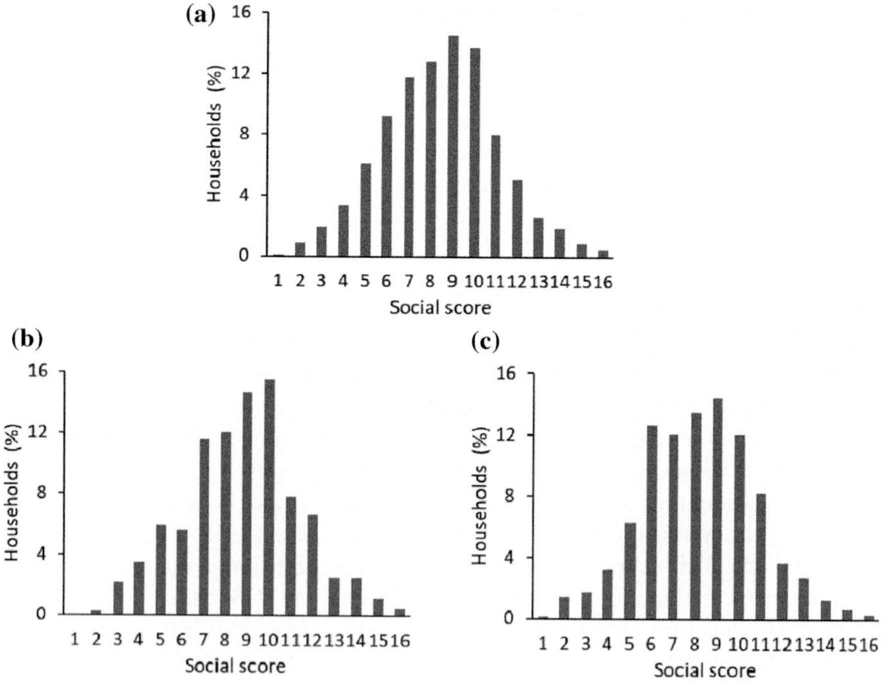

Fig. 1.4 Distribution of social scores **a** among all surveyed households (n = 1230), **b** among households in the N-transect, and **c** in the S-transect of Bengaluru, India

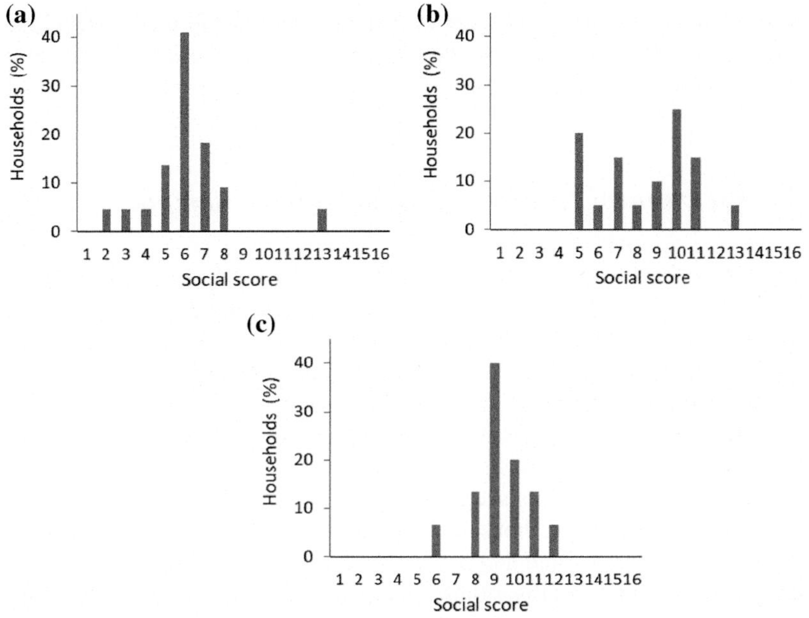

Fig. 1.5 Distribution of social scores in exemplary villages **a** Muninagara, **b** Bettarayanadoddi, **c** Doddakalsandra in the rural–urban interface of Bengaluru, India

were aggregated by stratum (Fig. 1.6). Apart from Sahakar Nagar, the wealthiest neighbourhood in stratum 1 of the N-transect, the highest socio-economic status seemed to prevail in stratum 3, which together with stratum 4 represents a transitional stage of urbanisation.

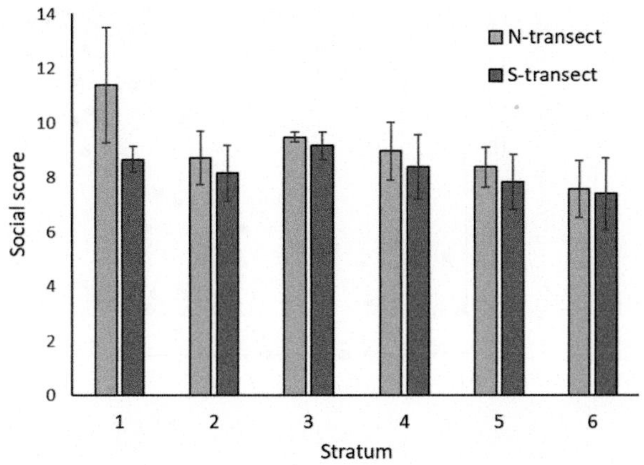

Fig. 1.6 Average social score of villages in the six strata along the rural–urban gradient in Bengaluru, India

Most of the papers in the subsequent chapters that refer to the household survey of FOR2432 used this delineation and categorised strata 1 + 2 as urban, 3 + 4 as transitional, and 5 + 6 as rural areas.

1.3.2 Agricultural Households in the Survey Sample

Engagement in farming proved to be another indicator for the degree of urbanisation in a given settlement. The proportion of farming households in a settlement was positively correlated with the SSI of the settlement, in particular in the transitional and rural areas of the transects, represented in strata 3–6 (Fig. 1.7).

The vast majority of farming households in these strata managed two to four plots, whereas high numbers of plots (>8) per farming household were only reported in the urban areas, where the share of farming households was low. Thus, with increasing urbanisation, agricultural plots seem to become fragmented and more laborious to manage. On the other hand, households tend to abandon agriculture, while those which stay with it may expand and potentially professionalise their farming activity.

GIS-based mapping visualises these characteristics in space (Fig. 1.8). While the share of agricultural households declined towards the more urbanised areas close to the city centre, social scores, and thus the average living standard of people tended to increase. At the same time, some villages did not follow this general trend and stand out, presumably due to local context factors or specific histories that remain to be explored (see Chap. 2). This kind of visualisation can thus be helpful

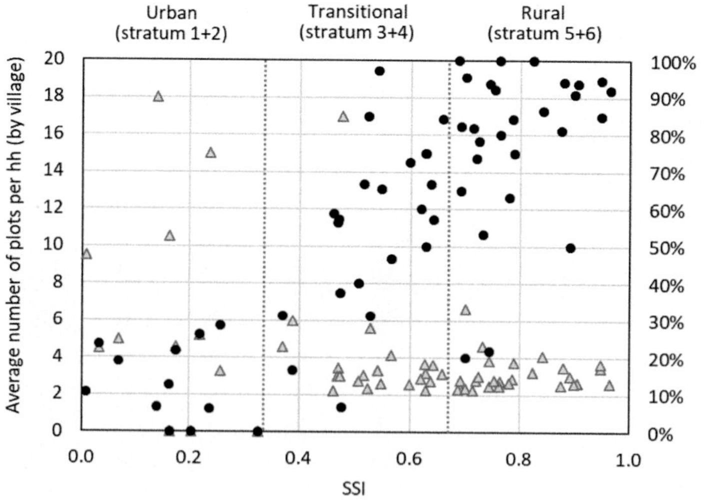

Fig. 1.7 Correlation analysis of agricultural activities and SSI. Round dark labels refer to the proportion of farming households per village, triangular grey labels indicate the average number of plots per household in the rural–urban interface of Bengaluru, India

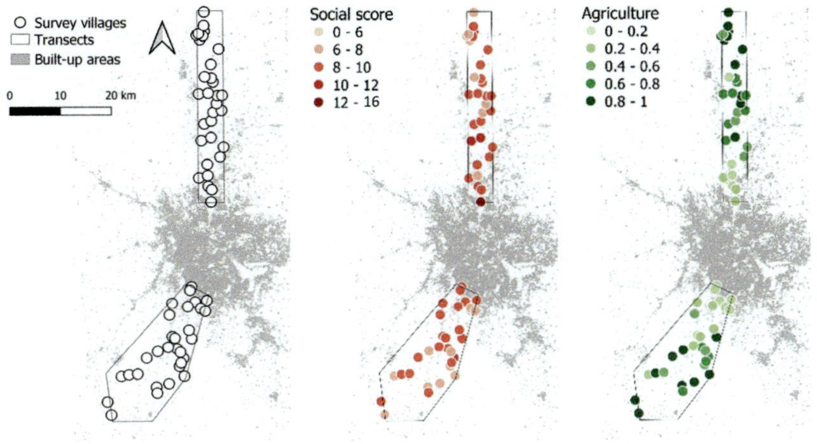

Fig. 1.8 Spatial distribution of average social score and share of agricultural households in the surveyed villages within the two research transects in Bengaluru, India

in both recognising common properties and patterns as well as spotting exceptions and hotspots.

1.3.3 Cropping Practice in the Survey Sample

In the questionnaire, usually addressed to the male household heads, all respondents were asked for the crops they cultivated on their plots in 2016, which may comprise one or more cropping seasons. Thus, in total 1938 records were taken, whereby only 29 plots were reported as fallow throughout the year. The names of the crop species were harmonised, and the frequency with which they were mentioned was evaluated to generate a 'profile' that described the preferences for and distribution of crops cultivated in the research transects.

Altogether (at least) 71 different crop species were identified in this survey. The crops were classified in eight groups (Fig. 1.9) as cereals, pulses, vegetables, fruits (incl. fruit trees), herbs and spices, flowers, grasses, and timber trees (incl. mulberry, thus termed 'trees and shrubs'). Cereals and pulses, as well as grasses and timber trees are groups of comparatively low diversity, but are widely cultivated. On the other hand, high species diversity was observed in vegetables, fruits, flowers, herbs, and spices.

At the species level (Fig. 1.10), the cereals grown by the surveyed households in the rural–urban interface of Bengaluru comprised the four species finger millet (ragi, *Eleusine coracana* Gaertn.), maize (*Zea mays* L.), rice (*Oryza sativa* L.), and sorghum (jowar, *Sorghum bicolor* (L.) Moench). Thereby the by far most important cereal was ragi (550 reports), followed by maize (352 reports). Maize is either cultivated

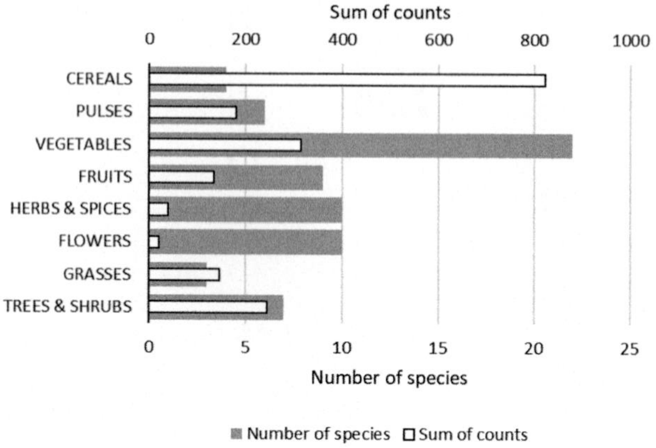

Fig. 1.9 Cropping profile by crop groups in Bengaluru, India

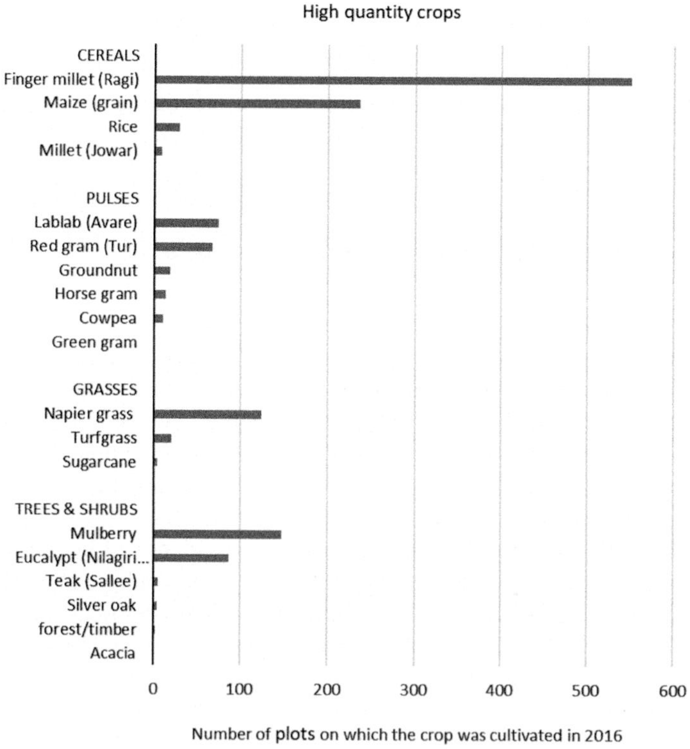

Fig. 1.10 Cropping profile within high-quantity crop groups in Bengaluru, India

to maturity for grain (236 reports), or harvested as baby corn (114, sweet corn 2) and thus marketed as a vegetable rather than a cereal. Rice (cultivated on 28 plots) and sorghum (8 plots) fell more than an order of magnitude behind. Altogether, 822 plots were used to cultivate some kind of cereal.

Pulses form an important part of the Indian diet, and six species were reported to be cultivated: Lablab (*Lablab purpureus* (L.) Sweet, mentioned 74 times), pigeon pea/red gram/tur (*Cajanus cajan* (L.) Millsp., 66), groundnut (*Arachis hypogaea* L., 18), horse gram (*Macrotyloma uniflorum* (Lam.) Verdc., 13) cowpea (*Vigna unguiculata* (L.) Walp., 10), and green gram (*Vigna radiata* (L.) R. Wilczek, 1). Altogether, 182 plots were allocated to grow some kind of pulses.

Among the other crop groups only Napier grass (*Cenchrus purpureum* (Schumach.) Morrone, 124) mulberry (*Morus* L., 147) and eucalyptus trees (nilagiri, *Eucalyptus* spp., 87) were mentioned more than 60 times. Mulberry is of high relevance due to the local silk industry, while Napier grass is an important fodder resource for Bengaluru's large livestock (dairy cow) population. Eucalyptus is often planted in plots designated for future sale to real estate, as it requires little maintenance and its wood can be used in construction works.

A high species diversity (Fig. 1.11) was encountered among the vegetables (22 species), fruits (9 species), herbs/spices, and flowers (10 species each). Apart from baby corn, the top vegetable species were tomato (*Solanum lycopersicum* L., 44 reports), green beans (*Phaseolus* spp., 35), and cabbage (*Brassica oleracea*, 18, including white cabbage and cauliflower), followed by chillies (*Capsicum* spp., 13), various kinds of gourds (*Cucurbita* spp.), and leafy vegetables such as spinach (*Spinacia oleracea* L.) or amaranth (*Amaranthus* L.). Eggplant (*Solanum melongena* L.) was reported three times. Fruit cultivation is dominated by coconut (*Cocos nucifera* L., 54 reports), grape (*Vitis vinifera* L., 28), mango (*Mangifera indica* L., 18), and banana (*Musa* spp., 15). With a high diversity of reported species, herbs, spices, and flowers feature as specialty crops, as only comparatively few farmers engaged in their cultivation.

The farmers were also asked about the size of the plots on which the respective crops were cultivated. If evaluated by cropped area the profile was similar. Just in the vegetable group, cabbage, potato, and cucumber would rank higher than spinach, chillies, and amaranth.

The two most important cereals (ragi and maize), the most important pulse (lablab), and three widely cultivated vegetables (tomato, cabbage, and eggplant) were chosen for a detailed investigation, how the crops respond to different agricultural management, and to characterise nutrient flows and use efficiencies under different intensity conditions in terms of nitrogen fertilisation and irrigation. The factorial field experiment setup for this purpose is described in Section II (Chaps. 7 and 8).

The plot sizes stated in the survey ranged from a few square metres (<0.1 ha) to 10 ha, with an overall average of 0.4 ha (Fig. 1.12). The vast majority of plots sized between 0.1 and 1.0 ha. The size distribution was identical in both research transects.

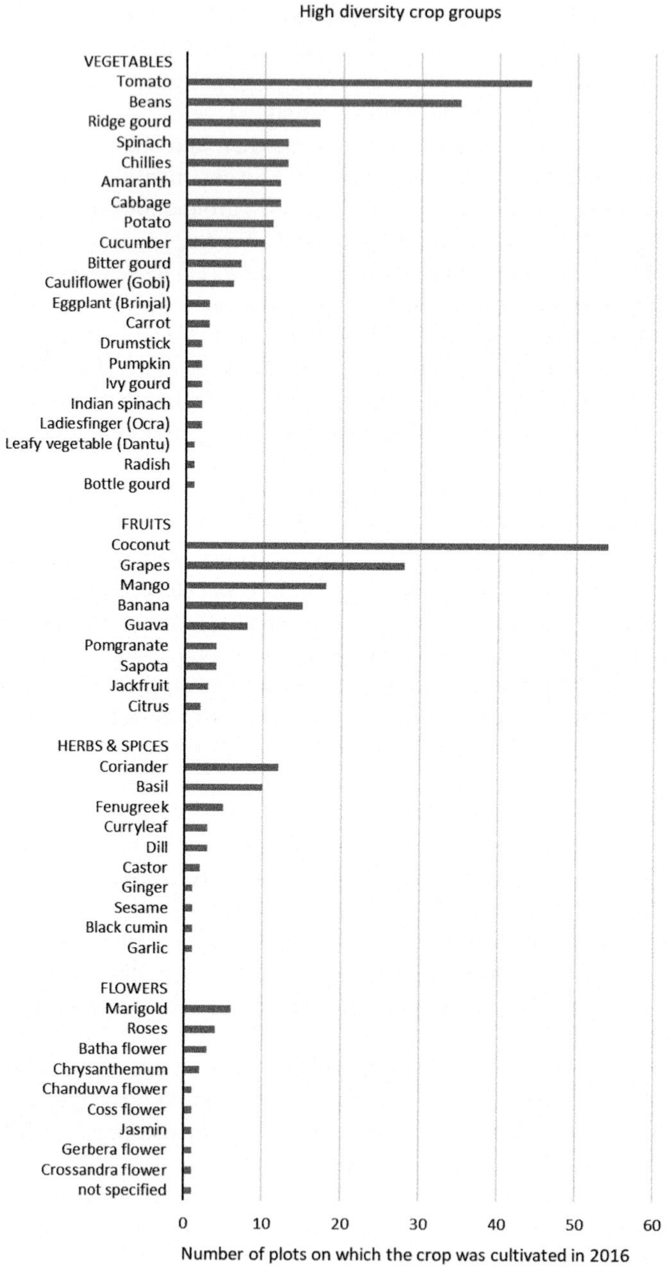

Fig. 1.11 Cropping profile within high diversity crop groups in Bengaluru, India

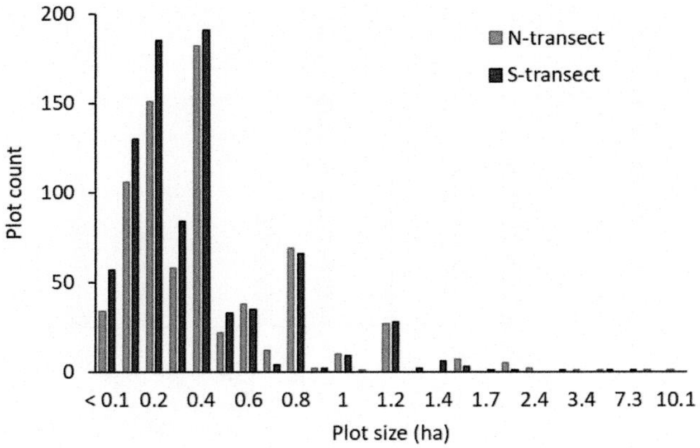

Fig. 1.12 Frequency distribution of plot sizes cultivated by the surveyed farmers in Bengaluru, India

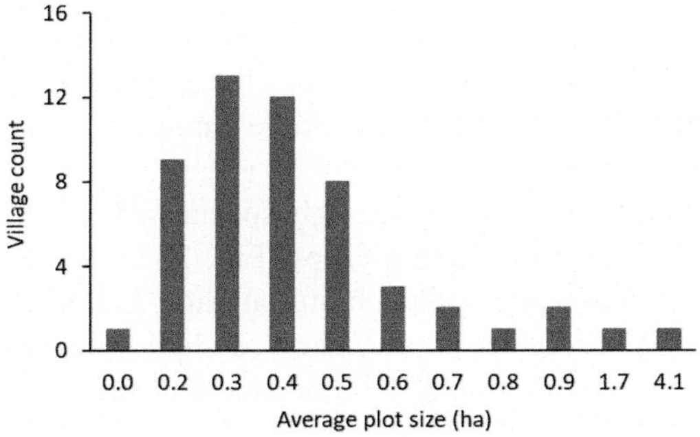

Fig. 1.13 Frequency distribution of average plot sizes cultivated in the surveyed villages in Bengaluru, India

The overall plot size distribution at the village level (Fig. 1.13) resembled that at the plot level (Fig. 1.12). The extreme values (very small and very large average plot sizes) occurred mostly in the urban strata or in very small rural villages, where only few data points were taken. There was no clear correlation to the SSI, and no significant differences between urban, transitional, and rural areas of the transects.

If analysed by crop group (Fig. 1.14), the plot size for fruit trees (0.63 ha) seemed to be above average (0.41 ha), that for herbs/spices (0.29 ha) and grasses (0.24 ha) below it; however, due to the wide range of sizes, this was not statistically significant.

The data presented in the cropping profile were also aggregated at the village level, in order to correlate them to the SSI. The results showed that the diversity of crops reported for a village increased with the SSI, i.e. towards more rural areas, but it also depended strongly on the size of the village given the number of households interviewed there, which ranged from 4 to 44. In 10 of the villages (mostly in the urban strata), the reports referred to less than 5 plots and are thus somewhat arbitrary.

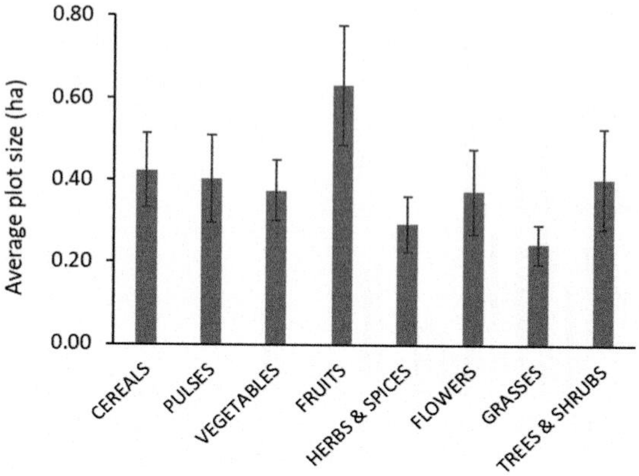

Fig. 1.14 Average plot size for different crop groups (error bars indicate variance) in the household survey in Bengaluru, India

A better measure for the crop diversity at the village or area level is the Herfindahl Index used in Chaps. 4 and 11, which combines the information on species numbers and the area on which they are grown in a single parameter.

In most villages, however, three or more crop groups were represented in a similar mix, while a few villages stood out as strongly specialised, as shown in the mapping of the distribution for some exemplary crops (Fig. 1.15). While some crops (such as Napier grass or tomato) are quite ubiquitous and evenly distributed over the transects, others (such as ragi or pulses) showed slight gradients along the rural–urban gradient, or were predominantly found in either the N-transect (eucalyptus, cabbage) or S-transect (coconut), or even in local hotspots (turf grass).

1.4 Conclusions

The distribution of socio-economic status and the prevalence of different crops along the research transects are just two features indicative for rural–urban transitions and agricultural practices. The examples given above show how data can be aggregated at different scales, and how mapping can reveal characteristic patterns. The subsequent chapters take this understanding as a starting point for specific questions and deeper analyses of drivers, outcomes, and interdependencies of various urbanisation-related processes. They characterise the rural–urban interface as a highly complex social-ecological system.

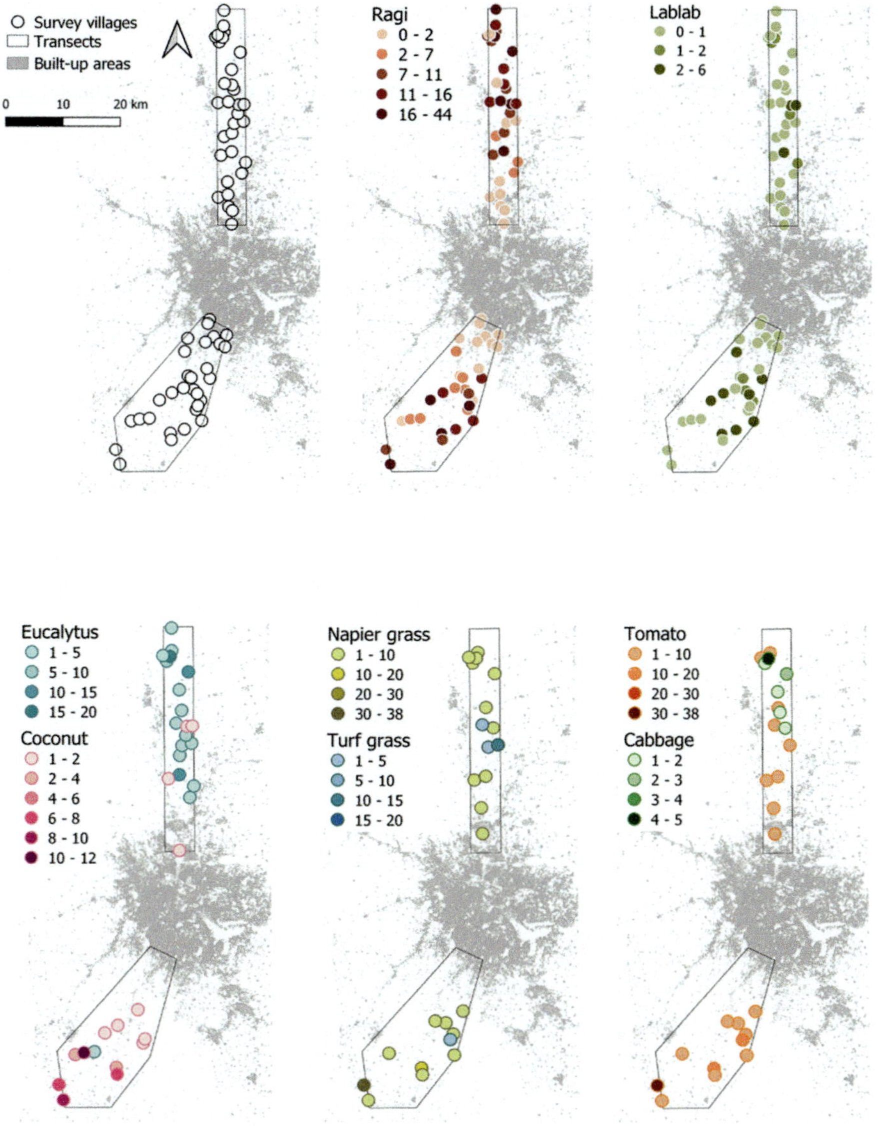

Fig. 1.15 Spatial distribution of exemplary crops, shown at the village scale in Bengaluru, India. Colour intensity of the labels increases with the frequency of cultivation in the surveyed villages

Acknowledgements The authors gratefully acknowledge the financial support provided by the German National Science Foundation, DFG, through grant number BU 1308/14-1, and by the Department of Biotechnology, DBT, India, through grant DBT/IN/German/DFG/14/BVCR/2016 as parts of the Indo-German consortium of DFG Research Unit FOR2432/1.

References

Hoffmann EM, Jose M, Nölke N et al (2017) Construction and use of a simple index of urbanisation in the rural–urban interface of Bangalore, India. Sustainability 9(11):2146

MRSI, Market Research Society of India (2011) Socio-economic classification 2011. The SEC system. http://imrbint.com/research/The-New-SEC-System-3rdMarch2011.pdf

World Population Prospects (2019) United Nations department of economic and social affairs. York, NY, USA

Part II
Changing Patterns of Land Use and Agriculture

Chapter 2
Temporal Dynamics of Land Use Change and Spatial Urbanisation Indices in the Rural–Urban Interface of Bengaluru Over the Past Three Decades

Sheetal Patil, Thanh Thi Nguyen, B. Dhanya, Seema Purushothaman, and Ellen Hoffmann

Abstract To capture the overall dynamics of urbanisation in the research space of FOR2432, this study tracks the land cover changes within the two transects across the rural–urban interface of Bengaluru. It covers a retrospective period of almost three decades, with observations in 1992, 2000, 2011, and 2016, based on Landsat and Sentinel satellite images, and aims at unravelling the dynamics of land use change as a prelude to deeper explorations of its drives. The spatio-temporal analysis of broad land cover classes relates it to the agricultural activities described in subsequent chapters of this volume, while for built-up land cover the focus lies on urban sprawl and the evolution of the Survey Stratification Index (SSI). The results underpin that the loss of vegetated and agricultural area increases both, along a spatial gradient towards the city, and over time. The spatial urbanisation pattern described by the SSI remains rather stable if data are aggregated over rural, transitional, and urban regions, or over six predefined strata, while individual villages show pronounced shifts in the SSI, that characterise them as nuclei of urbanisation. Characteristics of the "urban fringe" are highly variable between successive land cover classifications and have high dynamics of change.

S. Patil · B. Dhanya · S. Purushothaman
School of Development, Azim Premji University, Bengaluru 562125, India
e-mail: Sheetal.patil@apu.edu.in

B. Dhanya
e-mail: dhanya.b@apu.edu.in

S. Purushothaman
e-mail: seema.purushothaman@apu.edu.in

T. T. Nguyen · E. Hoffmann (✉)
Organic Plant Production and Agroecosystems Research in the Tropics and Subtropics, University of Kassel, 37213 Witzenhausen, Germany
e-mail: ellen.hoffmann@uni-kassel.de

T. T. Nguyen
e-mail: thanh.nguyen@uni-kassel.de

E. Hoffmann et al. (eds.), *The Rural-Urban Interface*, The Urban Book Series,
https://doi.org/10.1007/978-3-030-79972-4_2

Keywords Survey Stratification Index · Time series · Land use pattern · Land cover changes · Remote sensing · Urbanisation dynamics

2.1 Introduction

Many studies have documented urbanisation processes by the expansion of built-up areas around major cities (Taubenböck et al. 2008; Bagan and Yamagata 2014) and pointed out the implications of the associated loss of fertile land for agricultural production (Bren D'Amour et al. 2016). Based on remote sensing, they addressed scales from the global level (Seto et al. 2013) to exemplary metropolitan regions (Mohamed 2017), including Bengaluru (Taubenböck et al. 2008; Ramachandra et al. 2012; Kannan et al. 2020). While the general trend is undisputed, spatial heterogeneities become apparent in the more localised approaches aiming for higher spatial resolution (Ramachandra et al. 2012).

The research in FOR2432 confined itself to an even smaller area defined by two transects, covering 250 km^2 in the North (N-transect), and 300 km^2 in the South of Bengaluru (S-transect; Chap. 1). The transects comprise ca. 90 villages each, presumed to represent a rural–urban gradient not only in terms of land cover, but also in socio-economic household conditions, agricultural production systems, marketing and consumption practices (Chaps. 3–6, and Sect. 4). A specific analysis of land cover changes within these transects thus helps to contextualise other data taken at village or household scales. In itself, the land cover analysis here presents rural–urban transition dynamics at a spatio-temporal resolution unmatched by the majority of similar publications (Kannan et al. 2020).

The SSI as presented in Hoffmann et al. (2017) was a snapshot at a given time point (2016) which had allowed to position settlements on a conceptual urbanisation gradient, relative to each other (Chap. 1). Based on the percentage of built-up area within a fixed buffer area around a settlement centre, the actual SSI values would depend on the resolution of the satellite image, seasonal land cover changes, and classification accuracy, but the overall ranking at a given point in time should be fairly stable. Larger changes, however, would be expected in the retrospective time series, especially in the transitional areas where marked urban sprawl occurred.

A time series of land use changes within the FOR2432 transects was presented in Patil et al. (2018) as part of a broader social-ecological analysis. The present paper re-evaluates these data in more detail and combines them with the SSI approach, in order to gain a deeper understanding of the urbanisation dynamics in the rural–urban interface of Bengaluru during the past three decades.

2.2 Materials and Methods

2.2.1 Remote Sensing Data and Land Cover Classification

The land cover classification used in this time series analysis is described in detail in Patil et al. (2018). Landsat data (TM and ETM+, spatial resolution 30 m) were downloaded from the U.S. Geological Survey (USGS) and the Global Land Cover Facility (GLCF, http://www.landcover.org/). Two cloud-free pictures for the years 1992, 2000, and 2011 were obtained representing the local growing seasons, Rabi/dry season (March), and Kharif/rainy season (October/November/December). The analysis for 2016 was based on Sentinel data (spatial resolution 10 m). Quantum GIS (QGIS, http://www.qgis.org) was used for analysis. The image analyses included image registration, rectification and enhancement, false colour composite generation, and two-stage (unsupervised, followed by supervised) classification.

Land cover was originally assigned to eight classes (Fig. 2.1): The *Built-up* class consisted mainly of buildings (residential, commercial, industrial) and other sealed surfaces (transportation infrastructure) which result from human activities. *Waterbodies* described areas persistently covered with water, such as lakes, rivers, reservoirs, canals, and pools. *Vegetation* included areas with natural vegetation like shrubs, avenue trees, parks within the urban, whereas *Agriculture* included land cropped for production of food and fodder. Likewise, *Plantation* designated cultivated trees, such

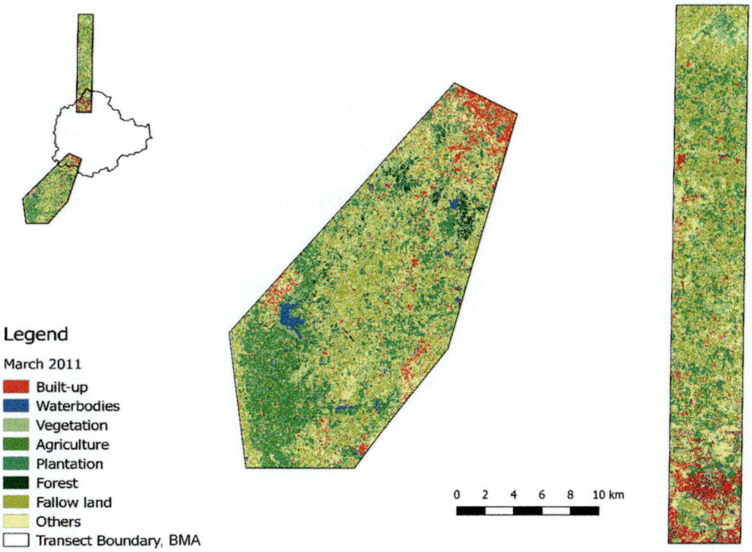

Fig. 2.1 Land cover classification within the research transects stretching out North and South of the Bengaluru Metropolitan Area (BMA), India, based on Landsat images; only one point within the time series is shown for illustration. For the full set of data refer to Patil et al. (2018)

as coconut, eucalyptus, mango, and other orchards, whereas *Forest* referred to natural tree cover (Bannerghatta National Park, reserve forests). *Fallow* consisted of non-vegetated cropland which is otherwise used for food or fodder. *Others* summarised natural bare soils, sand, or rocks and quarries, and anything not categorised in the above classes. For simplification in the temporal analysis, the classes agriculture, fallow and plantation were summarised as cultivated land, and the classes vegetation, waterbodies and forest as natural land, while built-up and others remained a separate class each.

2.2.2 *Survey Stratification Index (SSI) and Its Application in the Time Series Analysis*

The SSI developed in 2016 (Hoffmann et al. 2017), based on a Sentinel satellite image different from the one used by Patil et al. (2018), combines the percentage of built-up area within a 1 km^2 buffer area around a settlement/village centre with its linear distance to the centre of Bengaluru. With reference to the Bhuvan map, 93 and 98 villages were recorded in the N- and S-transect, respectively, and assigned an SSI ranging between 0 (urban) and 1 (rural). The numerical range was subdivided in six equal intervals (delineations at 0.167, 0.333, 0.5, 0.667, and 0.833), designated as *strata*, whereby stratum 1 would be the most urban, and stratum 6 the most rural one. Contours drawn around villages in the same stratum defined the strata areas used in the time series of land cover analysis (Patil et al. 2018). Their relative size ranged from 3 to 39% of the total transect area (Table 2.1).

For the present time series analysis of SSI evolution, only villages persistent throughout the years and buffers zones located inside the transect boundaries were included. The settlement coordinates and buffer zones defined in Hoffmann et al. (2017) for 2016 were overlaid to the land use maps generated by Patil et al. (2018) for 1992, 2000, 2011 (30 m resolution), and 2016 (10 m resolution), and the SSI was

Table 2.1 Relative size of strata within the transects and number of villages per stratum considered for the SSI time series analysis of Bengaluru, India

	Area %		Villages	
Stratum	N	S	N	S
1	6.1	2.6	5	14
2	13.0	3.3	8	10
3	9.3	16.1	8	14
4	19.8	1.3	17	28
5	29.3	39.4	28	24
6	22.5	19.4	22	8
Total	200 km^2	300 km^2	88	98

recalculated for all villages in each of the sampled time points. Thus, a set of seven SSI values was available for each village, in addition to the reference value from the original SSI analysis. The two independent analyses were assessed for consistency of the actual SSI values and the assignment of villages to the strata. A statistical t-test supported that differences between the seasons in the same year were non-significant, except for the data series of the N-transect in Nov 2000, which was therefore excluded from further evaluation. The graphical analysis below pertains to the March dataset. If single outliers within this dataset could be traced to misclassifications at another observation point, the respective village was omitted from the graph. In the N-transect this applied to two villages in stratum 2, three in stratum 3, three in stratum 5, two in stratum 6, and none in strata 1 and 4. In the S-transect, four villages were omitted in stratum 1, two in stratum 2 and 5, and none in the remaining strata 3, 4, and 6.

2.3 Results

2.3.1 Distribution of Land Cover Classes Along the Transects in 2016

Based on the land cover analysis for 2016, the distribution of land use classes along the two transects confirmed the concept of a rural–urban gradient (Fig. 2.2). In both transects, the built-up area occupied high shares in strata 1 and 2 which are close to Bengaluru city and declined steadily towards the more distant strata, while vegetated and agricultural areas showed the opposite trend.

Overall, the N-transect had more fallow land, whereas the S-transect had more natural vegetation and forests, but also a higher share of built-up area than the North in all of the strata. Waterbodies were sparse in the N-transect, except for the more urban strata (especially stratum 2), but were present in all strata of the S-transect.

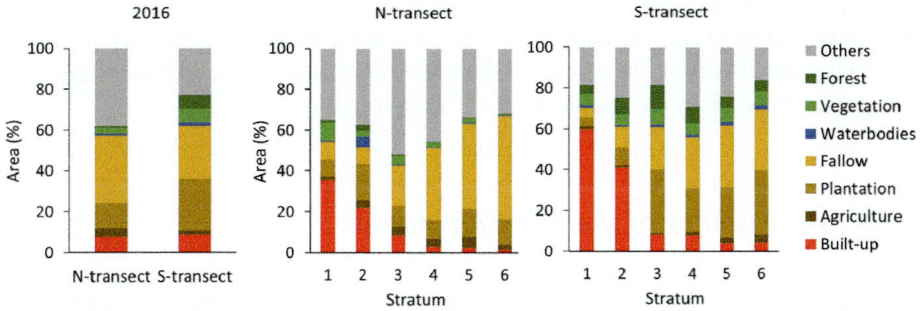

Fig. 2.2 Spatial distribution of land cover within two transects across the rural–urban interface of Bengaluru, India, in 2016

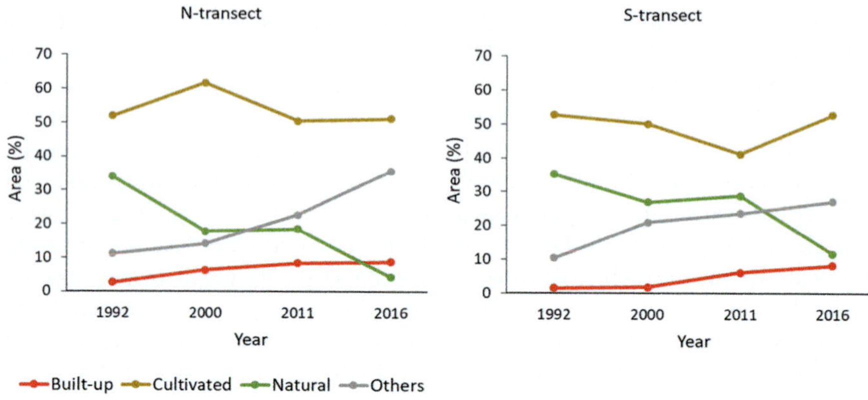

Fig. 2.3 Temporal changes of land cover within two transects across the rural–urban interface of Bengaluru, India, between 1992 and 2016

2.3.2 Land Cover Dynamics Over the Past Three Decades

To visualise the land use change over time, the land cover classes were summarised to four categories (built-up, cultivated, natural, and others) and evaluated for the entire transects (Fig. 2.3). Because the size of the strata differed markedly, the actual areas (in km^2) were added up for each land cover class over all strata, averaged between the two seasons, and then converted to relative values; this avoided over-weighting the small strata. Even at this highly aggregated scale, the increase in built-up area (fourfold, from ca. 2 to ca. 8%) and other land uses (threefold, from ca. 10 to ca. 30%) at the expense of vegetated area was evident, whereby natural vegetation was affected stronger than agricultural land. The increase in built-up area provides the background for the in-depth analysis of the SSI dynamics.

Compared to the overall (transect-wide) development, the past changes were much more profound in strata 1 and 2, which were already considered urban in the spatial gradient in 2016. In stratum 2, the share of built-up area increased from ca. 10 to 60% in the South (Fig. 2.4), and from 8 to 35% in the North (not shown). If plotted in a backward direction, the temporal dynamic of all analysed land cover classes in this stratum resembles the spatial gradient shown in Fig. 2.2.

2.3.3 Consistency of the SSI Calculation

The land use classifications of 2016 were based on Sentinel pictures with 10 m resolution, whereas the time series covering the points from 1992 to 2011 relied on Landsat images with 30 m resolution. The classifications, done independently by different research groups, distinguished five classes (built-up, vegetation, fallow, water, and others) in Hoffmann et al. (2017), and eight different land cover classes

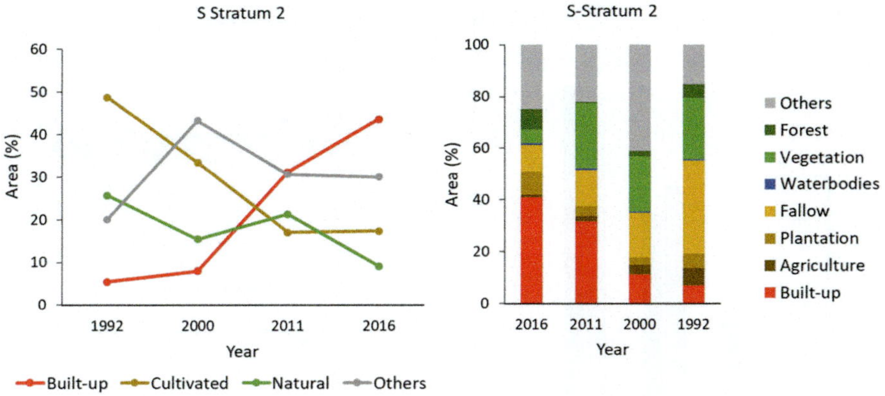

Fig. 2.4 Temporal changes of land cover in stratum 2 of the S-transect of Bengaluru, India, between 1992 and 2017, and exemplary visualisation of a "space-for-time substitution"

in the time series (see above). For the SSI calculation only built-up and non-built-up area matters. In a comparison of the SSI values calculated in both approaches for 2016, the number of villages assigned to urban, transitional, and rural strata was very similar (Fig. 2.5a).

With regard to the individual villages' SSI values, the SSI reference data were generally slightly lower than the results of the new analysis (Fig. 2.5b). Deviations between the two approaches (reference analysis of Hoffmann et al. 2017 and the new analysis presented here) were slightly higher than differences between the two seasons of the same year within the new analysis (Fig. 2.5c). The few exceptionally high deviations were traceable to misclassifications of land cover in the affected village at this observation point.

The strata boundaries were fixed values, and the assignments of villages to the strata were kept as delineated in 2016 based on the reference data, so that changes over time in the SSI of individual villages could be visualised as shifts within and between adjacent strata. Allocation mismatches in the most recent year are irrelevant when changes over time are analysed. In view of the data variability (Fig. 2.5), only consistent changes in SSI of more than 0.05 units over the entire period (that is between 1992 and 2016) were considered relevant.

2.3.4 *SSI Dynamics*

The average SSI in each stratum did not change much over the years. The biggest shift occurred in stratum 2 in the N-transect, and strata 1 and 2 in the S-transect, primarily between the years 2000 and 2011, but high standard deviations masked this effect. Changes of the average SSI in stratum 2 of the S-transect between the periods 1992–2000 and 2000–2011, and in stratum 5 and 6 of the N-transect between

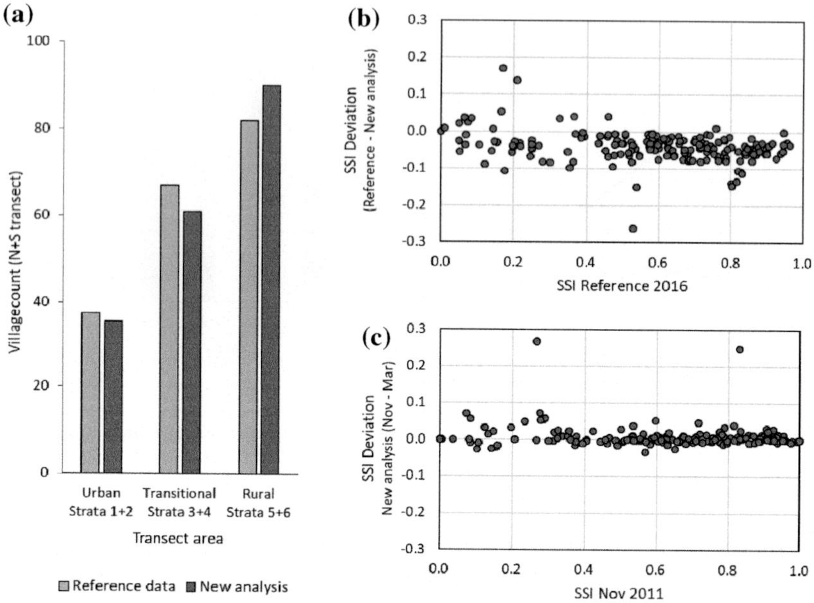

Fig. 2.5 Comparison of the SSI values calculated in the original analysis (Reference data according to Hoffmann et al. 2017) and in the new analysis based on the land use classifications of Patil et al. (2018). **a** Number of villages assigned to urban, transitional, and rural regions, **b** differences in single SSI values between the two analyses for 2016, and **c** differences between single SSI values between the two seasons (March and November) in 2011

1992 and 2000 were statistically significant. The overall spatial structure along the rural–urban transects was thus remarkably stable. This seemed to contradict the pronounced changes in overall land cover presented above. Clear changes in SSI, however, became apparent if data resolution was increased and SSI values were visualised for each individual village (Fig. 2.6). At that scale of analysis, also the strata showed different trends.

North transect

In stratum 1 (Fig. 2.6), there was overall little change, only for one village (Jakkuru) the SSI declined with time, such that it would have been in the neighbouring higher stratum until 2011. In stratum 2, the tendency for declining SSI values was evident and quite consistent, pointing to major urbanisation processes in this area. Three villages would have been allocated to a higher, more rural stratum until 2000 (Atturu, Anantapura) or 2011 (Venkatala). An overall slightly declining tendency was apparent in stratum 3. One village shifted from stratum 4 to the boundary of stratum 3 (Bettahalasur) and another village from stratum 3 to 2 (Manchihalli).

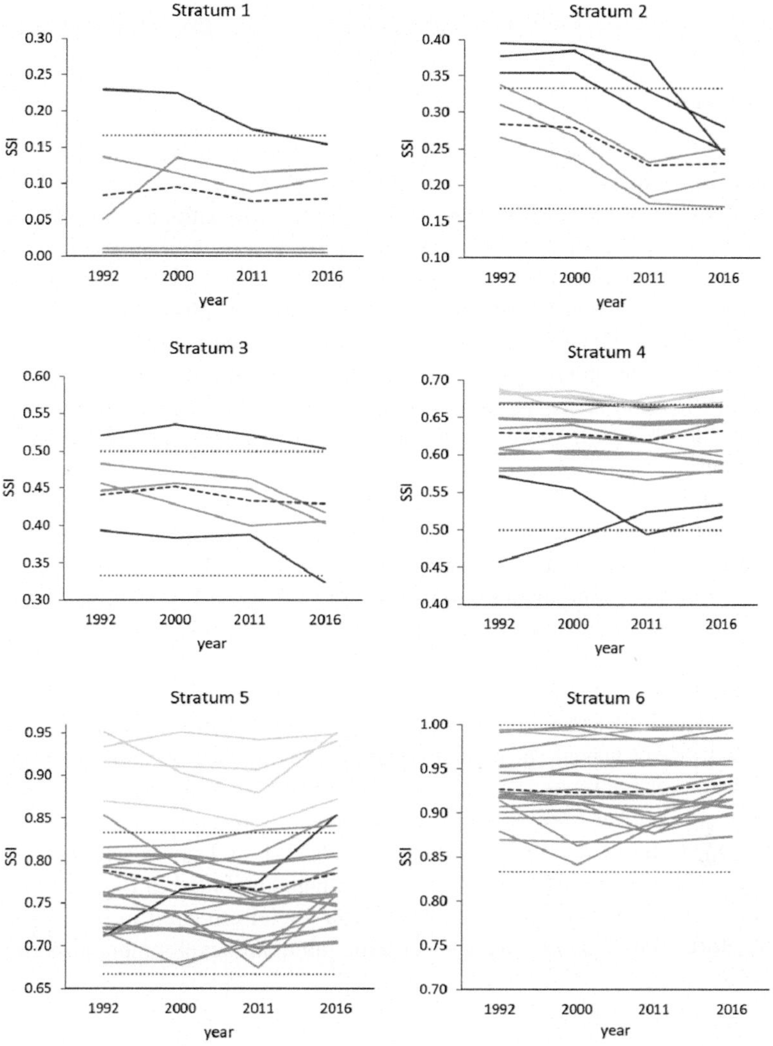

Fig. 2.6 Evolution of individual village SSI values for each stratum in the N-transect of Bengaluru, India. Dotted lines indicate the strata boundaries, the dashed line shows the average SSI for the stratum, and grey lines represent the individual villages. Consistent trends leading to a total change of more than 0.05 SSI units are highlighted in dark grey colour

The higher strata contained many more villages, and their SSI values changed less. According to the new analysis, five villages would have been allocated to stratum 5 rather than stratum 4 (with marginal differences), and four to stratum 6 rather than 5. In stratum 4, only two villages (of 22) showed a major change in SSI, one of them against the urbanisation trend (Addigandhalli and Nagadarsana-halli, respectively). In strata 5 and 6, there was overall little and inconsistent change, except for one village (Konaghatta) that gradually shifted from stratum 5 to the most rural stratum 6.

South transect

In stratum 1 (data not shown strata-wise), the SSI decreased over time for eight of the evaluated 10 villages, most marked between 2000 and 2011. A mismatch in strata allocation was noted for two villages in 2016, and three villages shifted newly into this stratum (Chuchgatta in 2000, Kodipur in 2011, and Subramanyapur in 2016). In stratum 2, all villages showed major decreases in SSI, and three had shifted into this stratum from the more rural neighbour (Talghatpur in 2011, Anjanpur and Ganiganpalya in 2016). The allocation of villages to strata 3 and 4 in 2016 showed some mismatches between the original and the present assessment. Using the Landsat-based classification, seven villages would belong to a higher, more rural stratum. However, only three villages evaluated in stratum 3 showed a major decrease in SSI (Kaggalipura and Tataguni after 2011, Nagegandanapalya already between 2000 and 2011). Only Tataguni shifted to a lower stratum. In stratum 4 (with four allocation mismatches), three villages out of 27 showed dramatic decreases in SSI after 2011 (Harohalli, Somanahalli, and Saluhanase), whereas all others hardly changed over time. This was very similar to stratum 5, where three out of 23 villages decreased in SSI (Bairamangala, Ganteganadoddi, and Hosur). The SSI of all eight villages of stratum 6 was rather stable over time.

A summarising view (Fig. 2.7), focused on the transforming villages over all strata, revealed differences in the SSI dynamics between the two transects. Stratum 1 in the N-transect was already mostly stabilised before 1992, and a very dynamic "fringe" area was confined to strata 2 and 3. The transforming villages in the South were more spread out over the entire transect. Here, the urbanisation trends were unfolding in stratum 1 mostly between 1992 and 2011, and successively set on in the higher strata. The mapping (Fig. 2.8) of the villages included in Fig. 2.7 confirmed this impression. In the S-transect, urbanisation nuclei emerged in particular along the Bengaluru–Kanakapura road, thus creating an axial development pattern.

2.4 Discussion

While the focus of the present study is on the SSI dynamics, the retrospective analysis of other land use changes within the research transects links this study to survey-based results presented in other chapters of this volume. The overall land use distribution is in good agreement with the cropping profile generated from the FOR2432 survey (Chap. 1), and with the subsurveys presented in Chaps. 3, 4, 11, and 19, and supports the idea of integrating different types of analysis within the FOR2432 project.

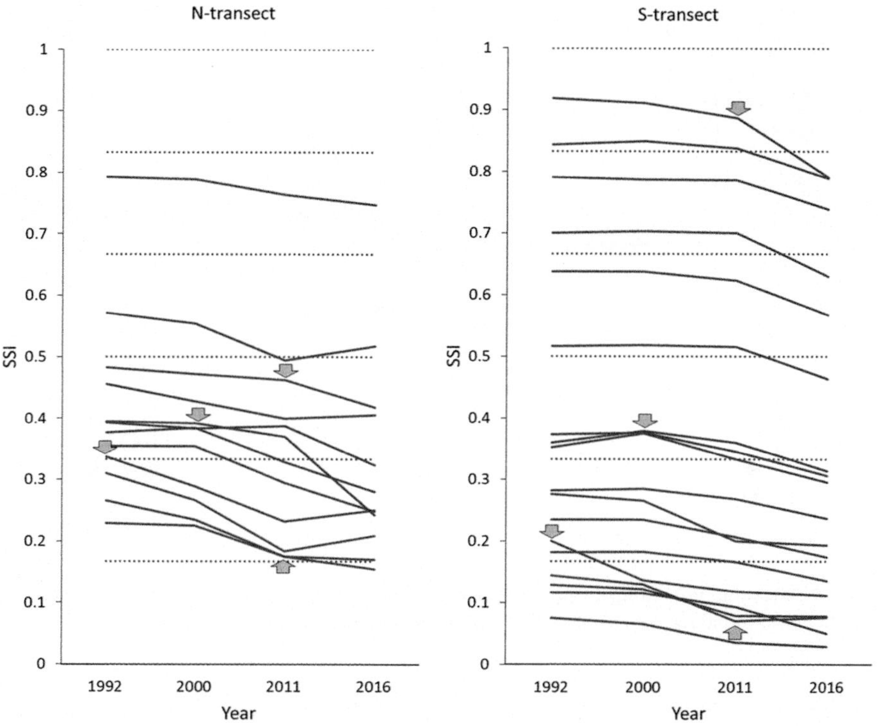

Fig. 2.7 Dynamics of SSI for villages with a decrease of more than 0.05 units between 1992 and 2016 in all strata of both transects across the rural–urban interface of Bengaluru, India. Downward arrows indicate the onset of an urbanisation trend for a group of villages, upward arrows a ceasing development

Changes in land use

One of the conceptual ideas behind the transect approach was that the conditions along a spatial gradient away from the urban centre might mimic the temporal development, such that more distant places resemble earlier stages of places closer to the city core. This has been proposed as "space-for-time substitution" and was applied mostly in ecological studies (Wogan and Wang 2018). Visualised in this way, the land use data supported this concept to some degree, especially for strata 1 and 2, which have experienced major urbanisation. The space-for-time approach, however, fails to capture socio-economic and political drivers of change (Damgaard 2019), which may be unique events, or changes in policy that redirect development from certain time points onwards, as well as spatial heterogeneities. For example, while Bengaluru is certainly the urban epicentre of the metropolitan area, smaller patches, or islands of built-up land also appeared within strata further away from the city. This indicates a more polycentric urban development pattern, in line with the study of Steinhübel and von Cramon-Taubadel (2020). Such urbanisation nuclei may also form around industrial estates or special economic zones, triggered by the master plans of city

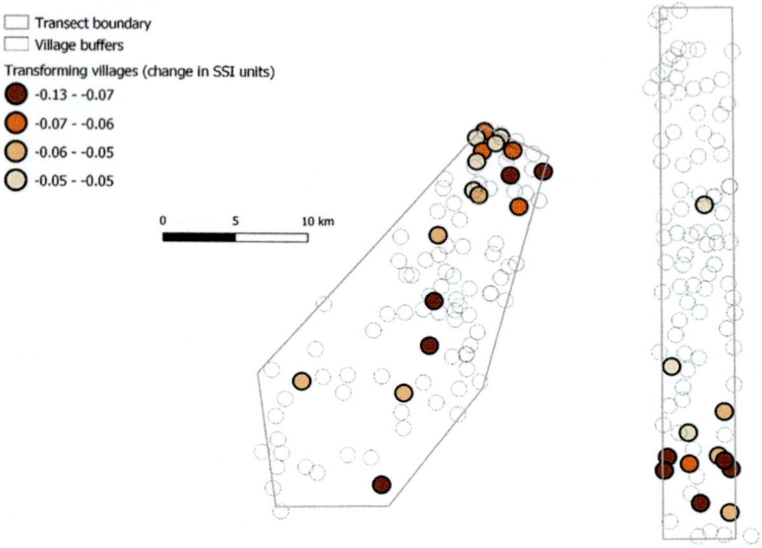

Fig. 2.8 Location of the transforming villages identified in Fig. 2.7 within the N- and S-transects of Bengaluru, India. Larger changes in SSI are indicated by darker colours

expansion. Nevertheless, these patches experienced their own local dynamics that either accelerated or curbed the development in response to receding ground water tables (Sekhar et al. 2018), labour moving to urban informal sector, low price of crop produce when input costs were high (Kannan et al. 2020), economic slow-down, over-supply in the number of apartments that came up from 2000s, and, later on, the currency denotification in 2016 (Purushothaman et al. 2016). Thus, when aiming at a social-ecological understanding, the space-for-time concept has strong limitations.

The definition of land cover classes is crucial for the explanatory power of the dynamics of change. The distinction of cropped area, fallow, and plantations within the cultivated land was justified by the objectives of the original study by Patil et al. (2018). Summarising crops and plantation may camouflage the increasing area under mango (*Mangifera indica* L.), eucalyptus (*Eucalyptus* spp.), and a few other tree species that replace the crop fields, and likewise, clubbing fallows with crops and plantations may conceal the conspicuous phenomenon of empty layouts around Bangalore (Hoffmann et al. 2017). These patterns may be useful to capture as the conversion within cultivated, agricultural land is strongly driven by Bangalore's expansion. While here this issue diverts too much from the actual focus, these trends are addressed in the analysis of cropping systems, such as in Chaps. 3 and 11 of the current volume.

SSI dynamics

The SSI analysis allows focusing on development at the village level. There were individual villages that stood out by continuous and relevant decreases in the index values over time, i.e. 12 in the North, and 18 villages in the S-transect. Their distribution in the strata, and onset or ceasing of transformations in certain groups of villages indicated that development in the N-transect was somewhat ahead of the South, but also more confined to a narrow transitional area. In the S-transect, on the other hand, there were several pockets of urbanisation, also further away from Bengaluru city, and likely also influenced by neighbouring medium-sized towns, such as Kanakapura (Steinhübel and von Cramon-Taubadel 2020).

In retrospective, stratum 2 in the North, and strata 1 and 2 in the South have experienced the most profound rural–urban transition, both in terms of land use changes and SSI dynamics, but the processes have come to cease at present. It will be good to understand if there was any jurisdictional, administrative, socio-economic, or other change that influenced these strata more than others. In most of the studies presented in this volume, strata 1 and 2 were considered as urban area and no longer as strongly transforming. This makes sense, as here we looked at the past development, whereas the other studies analyse the present situation. In 2016, the neighbouring strata 3 (and 4) featured as transitional area in many of the analysed parameters. In particular, stratum 3 in the North was an outlier in various aspects: Though close to the urban centre, it had a higher share of area under agriculture, and good water availability as documented in Patil et al. (2018). In the more rural strata, changes in production practices (Chap. 3) and socio-economic set-up were detectable even before they materialised in physical land cover changes.

2.5 Conclusions

The present study underlines that the evidence for rural–urban transformation becomes stronger as the scale of analysis decreases. Transformations unfold at a very local scale, from where innovations propagate or diffuse to the surroundings. Since a village-level investigation would avoid the dilution of effects due to averaging with nearby, non-transforming villages, it will be worthwhile to monitor the future fate of the villages that stood out as nuclei of transformation.

Acknowledgements The authors gratefully acknowledge the financial support provided by the German National Science Foundation, DFG, through grant number BU 1308/14-1, and by the Department of Biotechnology, DBT, India, through grant DBT/IN/German/DFG/14/BVCR/2016 as parts of the Indo-German consortium of DFG Research Unit FOR2432/1.

References

Bagan H, Yamagata Y (2014) Land-cover change analysis in 50 global cities by using a combination of Landsat data and analysis of grid cells. Environ Res Lett 9(6):1–13

Bren d'Amour C, Reitsma F, Baiocchi G et al (2016) Future urban land expansion and implications for global croplands. Proc Natl Acad Sci USA 114(34):8939–8944

Damgaard C (2019) A critique of the space-for-time substitution practice in community ecology. Trends Ecol Evol 34(5):416–421

Hoffmann EM, Jose M, Nölke N et al (2017) Construction and use of a simple index of urbanisation in the rural–urban interface of Bangalore, India. Sustainability 9(11):2146

Kannan E, Balamurugan G, Narayanan S (2020) Spatial economic analysis of agricultural land use changes: a case of peri-urban Bangalore, India. J Asia Pac Econ 26:1–17

Mohamed AM (2017) Monitoring of temporal and spatial changes of land use and land cover in metropolitan regions through remote sensing and GIS. Nat Resour 8:353–369

Patil S, Bashkar D, Vanjari RS et al (2018) Urbanization and new agroecologies: the story of Bengaluru's peripheries. Econ Pol Weekly 53(41):71–79

Purushothaman S, Patil S, Lodha S (2016) Social and environmental transformation in the Indian peri-urban interface—Emerging questions. Azim Premji University (Working Paper No. 1). https://www.researchgate.net/publication/298670759_Social_and_Environmental_Transformation_in_the_Indian_Peri-Urban_Interface_Emerging_Questions. Accessed on 03 Mar 2021

Ramachandra TV, Aithal BH, Sanna DD (2012) Insights to urban dynamics through landscape spatial pattern analysis. Int J Appl Earth Obs Geoinf 18:329–343

Sekhar M, Tomer SK, Thiyaku S, Giriraj P, Murthy S, Mehta VK (2018) Groundwater level dynamics in Bangalore City, India. Sustainability 10:26

Seto KC, Parnell S, Elmqvist T (2013) A global outlook on urbanization. In: Elmqvist T, Fragkias M, Goodness J et al (eds) Urbanization, biodiversity and ecosystem services: challenges and opportunities. Springer, Dordrecht, pp 1–12

Steinhübel L, von Cramon-Taubadel S (2020) Somewhere in between towns, markets and jobs—Agricultural intensification in the rural-urban interface. J Dev Stud. https://doi.org/10.1080/00220388.2020.1806244

Taubenböck H, Wegmann M , Berger C et al (2008) Spatiotemporal analysis of Indian mega cities. In: Proceedings of the international archives of the photogrammetry, remote sensing and spatial information sciences (ISPRS), Beijing, 3–11 July 2008

Wogan GOU, Wang IJ (2018) The value of space-for-time substitution for studying fine-scale microevolutionary processes. Ecography 41(9):1456–1468

Chapter 3
Urbanisation Effects on Agricultural Production Systems Across the Rural–Urban Interface of Bengaluru

C. L. Nandini, C. Kavya, G. N. Nagaraja, and P. S. Prasannakumar

Abstract Numerous studies address urbanisation effects on the redistribution of population and change in the demographic balance between rural and urban areas. Urbanisation, however, also strongly affects agriculture in the urban fringe where it changes cropping patterns, cropping sequence and cropping intensity. One of the assumed changes is a shift in cropping systems from the production of staple crops to vegetables. This hypothesis was tested in the rural–urban interface north of Bengaluru, India, along a transect subdivided in rural, transitional and urban regions. From each region, 30 vegetable farmers and 20 non-vegetable farmers were surveyed, assessing the change of area cultivated with different crops over the past five years. There was an overall reduction in land under staple crops, plantation crops and forest crops, while the area under horticultural crops and fodder cultivation increased in rural, transitional and urban regions, especially among vegetable farmers. The changes in relative areas under different crop types showed an increasing specialisation, and an increasing predominance of vegetable cultivation, both within the spatial gradient towards the urban centre, and over time, until, in the urban region, horticulture seemed to be the last form of agriculture to persist.

Keywords Agricultural production systems · Cropping pattern (crop area/crop species) · Cropping intensity · Vegetable farmers · Farmer survey · Transition

C. L. Nandini (✉) · G. N. Nagaraja
Agricultural Marketing, Co-operation and Business Management,
University of Agricultural Sciences Bangalore, Bengaluru 560065, India
e-mail: nandiniclokesh@gmail.com

G. N. Nagaraja
e-mail: drgnnagaraja@gmail.com

C. Kavya · P. S. Prasannakumar
Agricultural Economics, University of Agricultural Sciences Bangalore, Bengaluru 560065, India
e-mail: kavyagowda89@gmail.com

P. S. Prasannakumar
e-mail: shiprasannamk415@gmail.com

E. Hoffmann et al. (eds.), *The Rural-Urban Interface*, The Urban Book Series,
https://doi.org/10.1007/978-3-030-79972-4_3

3.1 Introduction

Bengaluru in the Indian state of Karnataka is well known for its dynamic rural–urban transformation and is one of the fastest growing cities in the world. The urbanisation process has resulted in transitions in land use, land values, water resources, lifestyles and livelihood options around the city. Urbanisation has been enhanced in the northern periphery by developments such as the construction of the new International Airport, national highways, Hardware Park, Financial City Project and other industries. Hence, this study focussed on agricultural production systems in Bengaluru North as this region has attracted a lot of activities, while growth in the southern part is reaching saturation.

Urbanisation has an impact on cropping pattern, cropping sequence and cropping intensity in agriculture at the urban fringe (Jaysawal and Saha 2014). Urbanisation is the key factor for underpinning and catalysing changes in land use, land transactions, increased rural–urban migration and the overall urban agricultural land use (Magigi 2013). Agricultural lands are most affected by rapid urbanisation which tends to deprive farmers of arable land thereby reducing agricultural productivity (Francis et al. 2012). The Food and Agriculture Organization of the United Nations defines peri-urban agriculture as agriculture practiced within and around cities, which competes for resources such as land, water, energy and labour (FAO 2001). Agriculture in peri-urban areas includes a wide range of activities such as fisheries, poultry, goat and cattle rearing, dairy farming, horticulture, floriculture and arable farming. Peri-urban agriculture has to cope with the long-term impact of rise in land prices associated with reduced size of land holding for agriculture, as well as the short-term impact of rise in agricultural wages (Gowda et al. 2012). Ultimately, urbanisation not only influences the production systems, but also paves ways for different means of income for farming households, which indirectly influences the standard of living of transitional and rural dwellers.

The most striking feature of urban agriculture, which distinguishes it from rural agriculture, is that it is embedded in and interacting with the urban economic and ecological system. Access to urban markets for agricultural produce and for seasonal labour opens up the possibility of cultivating on a commercial basis high-value, highly perishable crops such as leafy vegetables, replacing storable crops such as cereals and pulses (Te Lintelo et al. 2001). Also in Bengaluru, the majority of nearby farmers grow vegetables as a major crop due to easy access to market, transportation facilities and farmers direct market (Raithara Santhe) in Yelahanka. On this background, the study aimed to analyse the impact of urbanisation on the prevalence of non-vegetable and vegetable-based agricultural production systems across the rural–urban interface north of Bengaluru.

3.2 Materials and Methods

The present study was conducted in the northern transect defined in FOR2432 and followed the strata-based delineation of rural, transitional and urban regions (Chap. 1). In order to distinguish the two different production systems, farmers were considered as vegetable farmers (VF) if in 2016, at least 25% out of their total operational land was utilised for vegetable production, and as non-vegetable farmers (NVF) if less than 25% of their land was dedicated to vegetable production. The sample frame consisted of 150 farmers, with 50 each in the rural, transitional and urban area. Among each 50 farmers, 30 were vegetable farmers and 20 were non-vegetable farmers. The survey was carried out in 2017 by personal interviews using a pre-tested questionnaire. The data pertained to land holdings under different crops in 2015–16 as compared to five years ago (2010–11, by recall). The area under different crops in 2015–16 and 2010–11 was collected to know the change in cropping pattern over the years. To distinguish different cropping systems, crops were grouped as follows: cereals, pulses and oilseeds were categorised as staple crops; vegetables, fruit crops and flowers were categorised as horticultural crops; fodder crops contained Napier grass (*Cenchrus purpureum* (Schumach.) Morrone), alfalfa (*Medicago sativa* L.) and native gross green biomass; mango (*Mangifera indica* L.), eucalyptus (*Eucalyptus* spp.) and coconut (*Cocos nucifera* L.) were summarised as plantation crops; forest crops contained other woody species cultivated for non-food uses such as castor (*Ricinus communis* L.), neem (*Azadirachta indica* A.Juss), pongame oil tree (*Millettia pinnata* (L.) Panigrahi) and *Gliricidia*. Descriptive statistical tools were used to quantify the actual and relative changes in average land holdings of different crops and evaluate the trends regarding production systems (VF *vs.* NVF) and location in the rural–urban gradient (rural *vs.* transitional *vs.* urban areas).

3.3 Results

Irrespective of the production system, the average farm holding (agricultural area used per farm) decreased from rural to transitional to urban locations (Fig. 3.1). Forest, fodder and plantation crops occupied a much lower share of land than staple and horticultural crops. The area under forest crops declined towards urban locations already in 2011; hence, there was not much difference over the five years until 2016. Within the selected study area, plantation crops were marginal throughout. The land under staple crops decreased over time and towards urban locations, while the horticultural crops showed the opposite trend and in the urban region continued to increase even despite the overall gross loss in agricultural area.

Within each region (rural, transitional, urban), the average farm holdings of each crop type, its relative share of farm holdings and change over the five year period were analysed separately for vegetable (VF) and non-vegetable (NVF) production systems (Tables 3.1, 3.2 and 3.3). Vegetable farmers were defined by using, at the

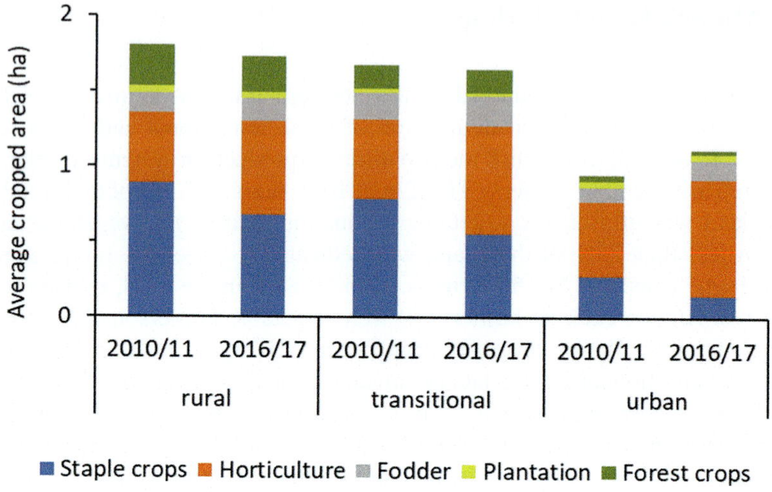

Fig. 3.1 Average farm holding for different crop groups of 50 selected farms (VF + NVF) per area along the rural–urban gradient of Bengaluru, India, in 2010/11 and 2016/17

Table 3.1 Production systems among vegetable vis-à-vis non-vegetable farmers in the rural region north of Bengaluru, India

Rural region		Average farm holding (ha)				Relative farm holding (%)				Change	
	Year	2011		2016		2011		2016		Difference 2016 – 2011 (%)	
	Production system	VF	NVF	VF	NVF	VF	NVF	VF	NVF	VF	NVF
Staple crops	Cereals	0.67	0.81	0.52	0.65	38.1	58.6	27.6	54.3	**−10.5**	−4.3
	Pulses	0.12	0.11	0.08	0.07	6.0	6.4	3.9	4.6	−2.1	−1.8
	Oilseeds	0.07	0.02	0.05	0.00	2.0	0.7	2.0	0.0	0.0	-0.7
Horticulture	Fruits	0.13	0.10	0.18	0.16	5.8	3.9	9.0	7.6	3.2	3.7
	Vegetables	0.42	0.10	0.61	0.01	25.2	6.0	34.3	0.3	**9.1**	**−5.7**
	Flowers	0.03	0.08	0.05	0.12	1.9	5.4	3.8	10.0	1.9	4.6
Fodder crops		0.13	0.11	0.14	0.16	0.16	8.8	7.9	12.1	0.7	3.3
Plantation crops		0.06	0.04	0.05	0.04	0.04	3.3	2.6	3.4	−0.4	0.1
Forest crops		0.33	0.18	0.27	0.18	0.18	6.9	9.0	7.7	−1.8	0.8

time of the interview, at least 25% of their land holding for growing vegetables; but on average, they even dedicated 34–42% of their area to these crops, whereas that share was between 3 and 0.3% for the non-vegetable farmers.

Five years earlier, the vegetable farmers used only 17–37% of their land holdings for vegetable production, whereas the farms classified under the non-vegetable

Table 3.2 Production systems among vegetable vis-à-vis non-vegetable farmers in the transitional region north of Bengaluru, India

Transitional region		Average farm holding (ha)				Relative farm holding (%)				Change	
	Year	2011		2016		2011		2016		Difference 2016 – 2011 (%)	
	Production system	VF	NVF	VF	NVF	VF	NVF	VF	NVF	VF	NVF
Staple crops	Cereals	0.65	0.60	0.43	0.52	36.6	49.8	21.8	53.1	**−14.8**	3.3
	Pulses	0.16	0.10	0.10	0.05	9.1	6.9	4.8	3.6	−4.3	−3.3
	Oilseeds	0.04	0.00	0.03	0.00	2.4	0.0	1.0	0.0	−1.4	0.0
Horticulture	Fruits	0.31	0.18	0.22	0.27	13.8	8.9	10.4	13.0	−3.4	4.1
	Vegetables	0.30	0.21	0.69	0.02	17.5	15.1	40.0	0.5	**22.5**	**−14.6**
	Flowers	0.00	0.01	0.04	0.09	0.0	2.0	1.8	7.8	1.8	**5.8**
Fodder crops		0.22	0.11	0.24	0.13	13.2	10.0	13.1	14.4	−0.1	−0.1
Plantation crops		0.03	0.00	0.03	0.00	1.2	0.0	1.5	0.0	0.3	0.3
Forest crops		0.15	0.16	0.15	0.16	6.3	7.4	5.7	7.5	−0.6	−0.6

Table 3.3 Production systems among vegetable vis-à-vis non-vegetable farmers in the urban region north of Bengaluru, India

Urban region		Average farm holding (ha)				Relative farm holding (%)				Change	
	Year	2011		2016		2011		2016		Difference 2016 – 2011 (%)	
	Production system	VF	NVF	VF	NVF	VF	NVF	VF	NVF	VF	NVF
Staple crops	Cereals	0.25	0.13	0.11	0.08	22.7	24.0	10.0	10.8	**−12.6**	**−13.2**
	Pulses	0.04	0.05	0.02	0.02	4.0	6.9	1.8	3.4	−2.1	−3.4
	Oilseeds	0.03	0.03	0.04	0.01	1.6	3.1	3.3	0.9	1.7	−2.2
Horticulture	Fruits	0.18	0.18	0.34	0.29	11.1	15.1	18.1	27.2	**7.0**	**12.2**
	Vegetables	0.32	0.11	0.50	0.02	37.0	9.6	42.1	3.4	**5.1**	**−6.2**
	Flowers	0.06	0.09	0.10	0.17	5.1	9.9	9.2	20.4	4.0	**10.5**
Fodder crops		0.11	0.09	0.12	0.15	11.3	23.4	12.6	24.8	1.2	1.4
Plantation crops		0.04	0.03	0.04	0.03	5.2	3.2	2.1	2.4	−3.0	−0.8
Forest crops		0.03	0.04	0.02	0.04	2.0	4.9	0.7	6.6	−1.3	1.7

production system had grown vegetables on 6–15% of their area. This points to increasing specialisation of the farming systems over time, as well as towards the city.

Among vegetable farmers across all three regions, oilseeds and plantation crops had a marginal share throughout, and the proportion of farm holdings of cereals, pulses and forest crops decreased as compared to five years ago (Tables 3.1, 3.2 and 3.3). The reduction was most pronounced for cereals, while the increase in relative farm holdings was most pronounced for vegetables and, in the urban region, fruits. Among the non-vegetable farmers, it was the proportion of farm holdings used for vegetables that decreased most pronounced over time in all regions. The change in relative area dedicated to cereals was small in the rural and transitional region, but declined sharply in the urban region. Here, the NVF farmers seemed to convert mostly to flowers and fruits. In both production systems, fodder cultivation increased slightly over time. For all major crop types, the largest changes were observed in the transitional region.

In both production systems, the differences of total and relative land holdings among the individual farm households were large, as were the measures of variance in the reported values. However, at the household level, there were 19, 22 and 11 farms classified as VF in 2016 that would *not* have classified as VF in 2011 in rural, transitional and urban areas, respectively; they shifted their production system from NVF to VF in the past five years. Vice versa, among the NVF in 2016, there were 2, 7 and 4 farms, respectively, that used more than 25% of their area for vegetables in 2011 and thus made a reverse transition (from VF to NVF). The four shifters in the urban region all converted to fruits and/or flowers and thus stayed with other horticultural crops, whereas this applied to only three of the remaining nine shifters in the other areas. The overall trend in system shifts is thus consistent with and validates the observed changes in actual and relative average land holdings for the respective crop types.

3.4 Discussion

The growth of the horticulture sector in Karnataka during the post reform period from 1992/93 to 2010/11 was analysed by Ramappa et al. (2015) at the state and district scale. They showed gradual variations in the rate of change in area under horticultural crops in an overall positive direction, as compared to significant variations in staple crops such as pulses, cereals and oil seeds. Furthermore, the horticultural crops gradually declined in drought prone areas, and the contribution to the total horticultural production was higher from vegetable and fruits in spite of lower area. Urbanisation changes agriculture by reducing the area under field and traditional crops, or increasing the area under commercial crops, as induced by urban demands (Harishkumar and Reddy 2017).

Asamoah (2010), Iheke and Nto (2010), Motamed et al. (2014), Regmi (2014) and Iheke (2015) also reported decrease in farm land due to urbanisation and thus a negative relation with agricultural production and productivity. Reduction in relative land holdings of cereals, pulses, oilseeds, plantation crops and forest crops, but not of fruit crops, vegetables, fodder crops and flower crops in rural, transition and urban regions among vegetable growers implies that increased urbanisation would lead to decrease in agricultural production of cereals, pulses, oilseeds, plantation crops and forest crops.

A study conducted by Gowda et al. (2012) in Magadi, in the western outskirts of Bengaluru, distinguished farms with high, low and no urban influence. They also found that the area devoted to each crop was relatively small in urban compared to rural farms, as the overall size of holdings is small. The average size of urban holdings (0.12 ha) was 85% smaller than rural holdings under low urban influence and 66% smaller than peri-urban holdings. Under high urban influence, 20% of the gross cropped area was devoted to ragi, 56% to vegetables (radish (*Raphanus sativus* L.) 16%, turnip cabbage knol-khol (*Brassica oleracea* L.) 14%, green leafy vegetables 12%, carrot (*Daucus carota* subsp. *sativus* (Hoffm.) Schübl. & G. Martens) and bean (*Phaseolus* spp.) 7% each), and 7% to flowers (China aster, *Callistephus chinensis* (L.) Nees).

In our study, reduction in the area under cereals and pulses was more pronounced among vegetable farmers compared to non-vegetable farmers. This may be related to the different water needs of the two production systems. On one hand, the majority of the cereals and pulses in the study area are grown under rainfed condition, but rainfall in the last three years was below normal in the study area, and crop failure was a common phenomenon. The vegetable farmers, on the other hand, knowing about the comparative profitability of growing vegetables than other crops, may have invested in irrigation facilities (Chaps. 4 and 6).

Further, decrease in area under vegetable cultivation by non-vegetable farmers can be attributed to drying of bore wells in the rural area and scarcity of water in transition and urban areas. In the majority of cases, farmers who discontinued vegetable cultivation left their land fallow or turned to dryland crops. Harishkumar and Reddy (2017) also highlighted that the area under water intensive vegetables and mulberry had decreased in all parts of the gradient (rural, transitional and urban) due to acute shortage of irrigation water, as the ground water is depleting at a rapid rate in Bengaluru North. Plantation crops, in this sample mainly mango and eucalyptus, are of minor importance in the study area, and even the meagre area under plantation crops decreased further, mainly because of non-economical productivity. For example, in eucalyptus plantations, the crop will come only once in 10 years, and in case of mango, the income is very low compared to other plantation crops.

Where heavy investments were made, primarily among the vegetable farmers, in the digging of bore wells in the dryland, the proportion of land holdings increased for flower crops, vegetables and fruit crops. This was enabled by electrification in the study area, but at the same time, the development of urban infrastructure forced farmers to shift to high-value horticultural crops to get better income and persist.

Thus, the cultivation practice by farmers in the urban area was demand driven. This is well in line with Te Lintelo et al. (2001) who showed that the increase in the share of land used for vegetables was partly explained by proximity to the markets. Another similar study (Raju 2011) revealed that the return to cost ratio was highest in flowers (3.3) followed by vegetables (2.0), plantation (1.9) and staple crops (1.6). Food production from peri-urban farmers has not been adequate to meet their consumption needs indicating the weakening of household food security of peri-urban farmers (Chaps. 12 and 13), unless it is compensated by higher income which allows them to purchase the food items they no longer produce themselves.

3.5 Conclusions

The increasing demand for local/regional food in urban communities and the role of direct marketing in that trend is strengthening peri-urban agriculture. There is a wide scope for increasing recognition of sustainable local/urban agriculture's role and capacity to contribute to urban ecosystems. This includes resource recycling and an increasing perception that there is a positive relationship between sustainable local/urban agriculture and the public health of associated communities. In India, smallholders around cities, towns and villages produce the majority of local food. In view of globalism and the increasing evidence that small-scale crop production is mostly incompatible with the purchasing mechanisms of major supermarkets, direct marketing is a local/regional response with the potential to strengthen smallholder agriculture.

Acknowledgements The authors express gratitude to Department of Biotechnology (DBT), Ministry of Science of Technology, Government of India for funding the Indo-German Collaborative research project "The Rural-Urban Interface of Bangalore: A Space of Transitions in Agriculture Economics and Society", and sub-project "Effects of urbanisation on value chains and livelihoods of farmers and other stake holders" executed by the University of Agricultural Sciences Bangalore, GKVK, Bengaluru (Sanction order no. BT/IN/German/DFG/14/BVCR/2016, GoI).

References

Asamoah B (2010) Urbanization and changing patterns of urban land use in Ghana: policy and planning implications for residential land use in Kumasi. Dissertation, Kwame Nkrumah University of Science and Technology

FAO—Food and Agriculture Organization of the United Nations (2001) Urban and peri-urban agriculture. A briefing guide for the successful implementation of urban and peri-urban agriculture in developing countries and countries of transition. SPFS/DOC/27.8, Revision 2, Handbook series volume III of the special programme for food security. Rome, Italy. http://www.fao.org/docs/eims/upload/215253/briefing_guide.pdf

Francis CA, Hansen TE, Fox AA (2012) Farmland conversion to non-agricultural uses in the US and Canada: current impacts and concerns for the future. J Agric Sustain 10(1):8–24

Gowda UCR, Chandrakanth MG, Srikanthamurthy PS et al (2012) Economics of peri-urban agriculture-case of Magadi of Bengaluru. Econ Pol Weekly 47(24):75–80

Harishkumar V, Reddy BVC (2017) Impact of urbanization on land use pattern of rural-urban gradient of Bengaluru North: an economic analysis. Econ Aff 62(2):303–312

Iheke OR (2015) Effects of sustainable land management practices on agricultural production: the case of arable crop farmers in Abia State, Nigeria. Internal J Agric Rural Dev 11(2):39–44

Iheke OR, Nto PO (2010) Effect of population pressure/urbanization on the adoption of sustainable agricultural practices by farmers: a case study of South Eastern Nigeria. J Food Fibre Prod 3(1):543–549

Jaysawal N, Saha S (2014) Urbanization in India: an impact assessment. Int J Appl Soc 4(2):60–65

Magigi W (2013) Urbanization and its impacts to food systems and environmental sustainability in urban space: evidence from urban agriculture livelihoods in Dar es Salaam, Tanzania. J Environ Prot 4:1137–1148

Motamed MJ, Florax RJGM, Masters WA (2014) Agriculture, transportation and the timing of urbanization: global analysis at the grid cell level (Working Paper, TI Discussion Paper, no 14-002/VIII), Amsterdam, Tinbergen

Raju K (2011) Impact of peri-urban agriculture on household food security—an economic study in the proximity of Magadi Town, Ramanagara district, Karnataka. Dissertation, University of Agricultural Sciences, GKVK, Bangalore

Ramappa KB, Jyoti U, Nagaraju Y (2015) Growth of horticulture sector in Karnataka-post reform period. Indian J Econ Dev 11(3):661–672

Regmi A (2014) The push-pull effects of urbanization on agriculture. Paper presented at the Reviving drylands dialogue: bridging policy and research, CGIAR Consortium, held at the African Studies Centre, Netherlands, June 17, 2014

Te Lintelo D, Marshall F, Bhupal DS (2001) Peri-urban agriculture in Delhi. Food Nutr Agric 29:4–13

Chapter 4
Changes in Land Use Pattern Vis-a-Vis Urbanisation and Credit Across the Rural–Urban Interface of Bengaluru

M. S. Udaykumar, K. B. Umesh, and P. S. Srikantha Murthy

Abstract This study analyses the changes in land use pattern of farm households in relation to urbanisation and credit, based on a survey among 50 randomly selected farmers in rural, transitional (peri-urban) and urban areas of the research transect north of Bengaluru, India. Data on cropping patterns before and after availing credit, and the Herfindahl Index were used to analyse crop diversification across the rural–urban interface. Credit enhanced diversification in all three areas, whereby the effect was stronger in transitional and urban areas than in the rural area. After availing credit, rural farmers diversified from staple crops to vegetables, fruit crops and livestock, while transitional and urban farmers diversified from staple and vegetable crops to high-value fruit crops, livestock and poultry. The area under staple crops decreased by 18.4%, 32.5% and 41.5% in rural, peri-urban and urban areas, respectively, while area under vegetables, flowers and fruit crops increased by 7.9%, 9.9% and 32.8%. Urbanisation provided an incentive for shifting from food crops to high-value horticultural crops and for more diversification towards commercial crops due to better access to market information and new supply and demand chains. Credit provided to farmers in time and in the required quantity would thus reduce the risk of incurring losses. Increasing the awareness on cultivation of high-value crops can lead to improvement in farmers' welfare and help to assure their food and livelihood security.

Keywords Agricultural production systems · Farmer survey · Credit · Agricultural assets · Crop diversification · Income

---

M. S. Udaykumar (✉) · K. B. Umesh · P. S. S. Murthy
Agricultural Economics, University of Agricultural Sciences Bangalore , Bengaluru 560065, India
e-mail: msudaykumar94@gmail.com

K. B. Umesh
e-mail: umeshuas@gmail.com

E. Hoffmann et al. (eds.), *The Rural-Urban Interface*, The Urban Book Series,
https://doi.org/10.1007/978-3-030-79972-4_4

4.1 Introduction

Urbanisation is a worldwide phenomenon as it is a part and product of economic change. Today, more than half of the world's population lives in cities, and it is estimated that by 2030, two-third of the world's population will be urbanised (Kalamkar 2009). According to the Food and Agricultural Organization (FAO), by 2020 Africa, Asia and Latin America will host 75% of urban dwellers (Anonymous 2017). Urbanisation not only leads to changes in socio-economic conditions, but also to their ecological transformation (Grimm et al. 2008; Turok and McGranahan 2013). The rapid urbanisation across the globe poses a serious question on how to manage agro-ecosystems to feed the growing population in a sustainable way.

In India, the concentration of urban population has been increasing over the years as a result of migration from rural to urban areas (Thorat et al. 2011) leading to large changes in land utilisation pattern. In this phase of rapid urbanisation, access to credit for agriculture plays an important role to feed the growing population and to reduce the migration of households from the rural agricultural sector to urban areas in search of jobs. Credit is the life-blood of agricultural economy (Hartarska et al 2015). It plays a pivotal role in generating the livelihood of farm households as it affects the adoption of new technologies to raise agricultural productivity which in turn affects the income of the farm households. The credit bridges the gap between farm household's financial requirements for agricultural operations and the ability to cover the same from their own funds. Limited capital availability has been acknowledged as the most important constraint in the modernisation of agriculture (Dantwala 1966).

Changes in land use patterns are linked to increasing income of farm households which in turn increases the consumption of dairy products, meat and meat-based products (deHaen and Thompson 2003). Urbanisation usually goes along with economic development. Meeting the needs of a growing population with limited and decreasing agricultural land holding, however, requires increasing agricultural productivity by advanced technologies. Many farmers can only adopt those through better access to agricultural credit. In this context, this study analysed the interlinkages between urbanisation, land use pattern and credit along a rural–urban gradient in Bengaluru, India.

4.2 Data and Methodology

The study was conducted in the rural–urban interface of Bengaluru, Karnataka, during January and February, 2017. The northern transect as defined in FOR2432 was divided into rural, transitional (peri-urban) and urban areas, based on the Survey Stratification Index (Hoffmann et al. 2017). In the three areas, 20 villages were selected randomly, of which 18 coincided with the sample for the large household survey (Chap. 1). A purposive stratified random sampling method was then adopted for the selection of two to 10 borrower-farmer households within the villages, none of which overlapped with the other survey. The sample frame finally consisted of 150 farmers representing 50 each from the rural, transitional and urban areas.

The data were collected from the selected farmers by personal interview using a pre-tested, well-structured schedule developed for the study. The information elicited from the respondent farmers pertained to land holdings, asset position, family size, educational level and decision making. With regard to availing credit, the survey asked for the borrowed amount, the different sources of credit and the duration, distinguishing short-term and long-term loans; in case of long-term loans, only farmers who borrowed credit since 2013 were considered. The utilisation pattern of credits was explored with regard to cropping pattern and income from different sources before and after availing credit. Farmers' statements on financial returns from different crop types, their information sources and market participation were considered to interpret and contextualize the findings.

The present paper puts its emphasis on examining the impact of credit on farming practice, crop diversity and associated change in land use patterns. Credit taken from the regular banking sector (commercial banks, cooperatives and regional rural banks) was considered formal, while informal sources of credit would comprise money lenders, traders, merchants, commission agents, friends and relatives. Diversification was analysed from the perspective of the farm household (enterprise diversification) and from the spatial/agroecological perspective of how the available cropping area was used (crop diversification). Net cropped area was defined as the total area of crops and orchards in which crops are grown only once in a year. Gross cropped area described the total area sown with crops and orchards once or more in a year. The ratio of gross to net cropped area Eq. (4.1) is referred to as cropping intensity and describes the raising of several crops from the same field during one agricultural year.

$$\text{Cropping Intensity} = \text{Gross Cropped Area/Net Cropped Area} \times 100 \tag{4.1}$$

Overall crop diversity was analysed in detail using the Herfindahl Index (HI). It is the sum of the squared proportions of acreage under each crop in relation to the gross cropped area (Eq. 4.2), where P_i represents the acreage proportion of the ith crop in the total gross cropped area.

$$\text{HI} = \sum_{i=1}^{N} P_i^2 \tag{4.2}$$

The Herfindahl Index takes the value of 1 when there is specialisation and approaches zero when there is diversification and N is large. HI was calculated at the area level for all the three regions.

The crops reported in the survey were grouped as follows: Cereals and pulses are summarised as staple crops with a comparatively lower market value. Vegetables are a highly diverse group comprising more than 20 different species recorded in this survey. Flowers (*Rosa*, *Chrysanthemum*, *Gerbera*, *Tagestes*) and fruit crops (grapes

(*Vitis vinifera*), sapota (*Pouteria sapota*), guava (*Psidium guajava*), banana (*Musa*)) are high-value crops cultivated by farmers in the study area. Perennials comprised teak (*Tectona grandis*), silver oak (*Grevillea robusta*), eucalypt (*Eucalyptus*) and commercial lawn (turf grass). The animal husbandry activities considered in this study were livestock (cross bred cows, local cows, buffaloes, sheep, goat) and poultry.

4.3 Results and Discussion

4.3.1 Total Credit Borrowed by Farmers from Different Sources

The total quantum of credit borrowed by farmers was highest in the urban area, followed by transitional and rural regions (Table 4.1). The total amount borrowed from formal sources was higher than that from informal sources throughout, but the share of formal credit increased towards the city, with ca. 60, 76 and 84% of the total in rural, transitional and urban areas, respectively.

This can be attributed to better accessibility of formal banking services in urban and transitional areas, with reasonable rate of interest and security as compared to informal sources. It is interesting to notice that farmers in rural areas borrowed more from informal sources because of the timely availability and easier accessibility of funds in that area. Further research into this aspect addresses the transaction costs associated with the different credit sources in more detail (Udaykumar et al. 2020). In general, the credit needs of farmers have been growing due to increase in the commercialisation of agriculture across all the areas (Anonymous 2010).

Table 4.1 Amount and sources of credit across the rural–urban interface of Bengaluru, India. Average total credit per household in rupees and (in parentheses) share of formal and informal sources

Transect area	Formal sources	Informal sources	Total
Rural	178,000	119,000	297,000
	(60%)	(40%)	(100%)
Transitional	286,000	91,000	377,000
	(76%)	(24%)	(100%)
Urban	434,000	82,000	516,000
	(84%)	(16%)	(100%)

4.3.2 Enterprise and Crop Diversification by Farmers Across Rural Urban Interface

4.3.2.1 Enterprise Diversification

After availing the loan, the number of farmers cultivating staple crops decreased throughout the transect, while the number of farmers reporting other farming activities increased (Table 4.2). Rural farmers diversified from staple crops to vegetables, fruit crops and livestock, whereas transitional and urban farmers chose to grow high-value horticulture crops and diversified from staple and vegetable crops to fruit crops, livestock and poultry.

Growing staple crops was less profitable as compared to growing of horticultural and commercial crops. It was interesting to note that six farmers in the transitional area had taken up commercial lawn cultivation after availing credit. Before availing credit, one farmer in rural, seven farmers in transitional and 15 farmers in urban area were growing fruit crops. After availing credit, the number increased to 12, 19 and 28 farmers, respectively.

These horticultural crops have two to three times higher productivity per rupee of investment than staple crops and vegetables, and are considered 'low volume—high-value' crops requiring less agricultural land. Increased awareness on cultivation of high-value crops, increased access to market information and engagement in new supply and demand chains has allowed the farmers to fetch higher returns and fostered the shift from staple crops to high-value crops. This has led to improvement in the farmer's welfare, well in line with the results of Kumar and Gupta (2015).

With regard to the land use pattern, the area under staple crops decreased to an extent of 18%, 33% and 42% in rural, transitional and urban areas, respectively. On

Table 4.2 Enterprise diversification by farmers across the rural urban interface of Bengaluru, India. Number of farmers engaged in the activities and (in parentheses) the average cultivated area in ha or herd size

	Rural ($n = 50$)		Transitional ($n = 50$)		Urban ($n = 50$)	
Farming activities	Before availing credit	After availing credit	Before availing credit	After availing credit	Before availing credit	After availing credit
Staple crops	49 (0.95)	48 (0.77)	46 (1.13)	36 (0.76)	43 (1.00)	36 (0.59)
Vegetables	26 (0.54)	32 (0.67)	20 (0.42)	22 (0.49)	21 (0.53)	23 (0.63)
Flowers	3 (0.02)	6 (0.06)	5 (0.05)	5 (0.05)	5 (0.06)	6 (0.07)
Fruit crops	1 (0.01)	12 (0.16)	7 (0.07)	19 (0.45)	15 (0.17)	28 (1.05)
Perennials	4 (0.01)	3 (0.08)	5 (0.30)	5 (0.32)	6 (0.07)	7 (0.25)
Lawn	0	0	0	06 (0.06)	0	0
Livestock	35 (2.68)	46 (3.15)	20 (2.20)	31 (3.20)	26 (2.60)	33 (3.40)
Poultry	0	1	0	0	0	4

Table 4.3 Crop diversification by farmers across the rural urban interface of Bengaluru, India. Use of the agricultural area and (in parentheses) relative changes after availing credit

	Rural ($n = 50$)		Transitional ($n = 50$)		Urban ($n = 50$)	
Particulars	Before availing credit	After availing credit	Before availing credit	After availing credit	Before availing credit	After availing credit
Gross cropped area (ha)	1.61	1.73 (7.4)	1.98	2.14 (8.1)	1.85	2.60 (40.0)
Net cropped area (ha)	1.51		1.61		2.29	
Cropping intensity (%)	106.66	114.56 (7.9)	122.98	132.91 (9.9)	80.78	113.53 (32.8)
Herfindahl Index	0.49	0.37	0.40	0.25	0.34	0.29

the contrary, area under vegetables, flowers and fruit crops increased, as well as the overall diversity of cultivated crops. This is in good agreement with a general shift towards vegetable cropping systems as shown in Chap. 3.

4.3.2.2 Crop Diversification

The farmers achieved crop diversification by raising the cropping intensity, rather than by extending their land holdings. Net cropped area before and after availing credit remained the same with an average of 1.5, 1.6 and 2.3 ha per household across rural, transitional and urban areas (Table 4.3). At the same time, after availing credit, the gross cropped area increased by 7.4%, 8.1% and 40% in rural, transitional and urban areas, respectively. The increase in cropping intensity was thus to the extent of 7.9%, 9.9% and 32.8% in rural, transitional and urban areas, respectively.

Based on the land use data aggregated over all households in the rural, transitional and urban group, the HI indicated that greater diversification of crops was observed after availing credit in all the three areas. After availing credit, the extent of diversification was highest in transitional (0.25) areas, followed by urban (0.29) and rural (0.37) areas.

Farmers of transitional and urban areas had better access to information and technology than rural farmers. Urban and transitional farmers were more confident in reaching consumers with their produce. Peri-urban and urban farmers chose direct marketing to consumers to sell their produce which allowed them to fetch high income, and this was also the major reason for more diversification towards commercial crops. The values of HI (Table 4.3) revealed that agriculture was transforming from traditional subsistence agriculture to high-value agriculture (Kumar and Gupta 2015). As a result of urbanisation, high-value horticulture crops have reliable markets and a higher potential in increasing farm income. The results are similar to the study conducted by Birthal et al. (2007). Crop diversification is important to farmers to mitigate the risk on one hand, and on the other hand, diversification is necessary to meet the different needs and demands of the consumer. Having diversification is always beneficial in farming business. The higher the diversification, the better will be the profit and resilience. Providing credit access to farmers in time and in required quantum will assist in crop diversification efforts (Komarek 2010). The trends in crop diversification further support the shift from non-vegetable to vegetable production systems (Chap. 3).

4.4 Conclusions and Policy Recommendations

Agricultural credit is an important and valuable tool to assist the economic development of a country by means of empowering the farming community. The development of the agriculture sector is highly dependent on the banking sector because 80% of farmers are small and marginal and thus unable to save and invest due to their low income levels. The results highlighted that availability of credit is a major determinant of crop diversification. The crop diversification in turn has strong effect on increasing the farm income. Availability of credit facilitated the farmers to adopt high-value horticulture crops. Contrary to this, the area under staple crops had decreased which in turn may have negative impact on food security, if not compensated by higher income from horticultural crops. On the other hand, crop diversity was positively linked with dietary diversity and health status of the household members (Chap. 11). Availability of credit in all the three areas of the transect led to increased cropping intensity and greater crop diversification than before.

Our results suggest that with urbanisation, the area under food crops is declining. To avoid negative implications on the food security of households, it is suggested to strengthen integrated farming systems (Gill et al. 2009). Policies to ensure access to formal credit by smallholder farming households are also recommended. To improve further the dietary diversity, efforts should be made to increase household income from agriculture, as well as non-farm and off-farm sources.

Acknowledgements The paper is the part of the Indo-German Collaborative Research project titled "The Rural–Urban Interface of Bangalore—A Space of Transition in Agriculture, Economics and Society"—Sub project "Food Insecurity at Different Stages of Urbanisation" (Sanction order: BT/IN/German/DFG/14/BVCR/2016 dated 16th November 2016). The Department of Biotechnology (DBT), GOI, New Delhi is duly acknowledged for the financial support of this project.

References

Anonymous (2010) Report of task force on credit related issues of farmers, pp 1–83
Anonymous (2017) The future of food and agriculture, trends and challenges. Food and Agriculture Organization of the United Nations, Rome

Birthal PS, Joshi, PK, Roy D et al (2007) Diversification in Indian agriculture towards high-value crops. IFPRI discussion paper, pp 1–29
Dantwala ML (1966) Institutional credit in subsistence agriculture. Int J Agrar Aff 5(1):52–61
deHaen H, Thompson H (2003) Food security in a world without borders. Forum Nutr 56:26–32
Gill MS, Singh JP, Gangwar KS (2009) Integrated farming system and agriculture sustainability. Indian J Agron 54(2):128–139
Grimm NB, Foster D, Groffman P, Grove JM, Hopkinson CS, Nadelhoffer KJ, Peters DP (2008) The changing landscape: ecosystem responses to urbanization and pollution across climatic and societal gradients. Front Ecol Environ 6(5):264–272
Hartarska V, Nadolnyak D, Shen X (2015) Agricultural credit and economic growth in rural areas. Agric Financ Rev 75(3):302–312
Hoffmann EM, Jose M, Nölke N et al (2017) Construction and use of a simple index of urbanisation in the rural–urban interface of Bangalore, India. Sustainability 9(11):1–21
Kalamkar SS (2009) Urbanization and agricultural growth in India. Indian J Agric Econ 64(3):1–20
Komarek AM (2010) Crop diversification decisions: the case of vanilla in Uganda. Q J Int Agric 49(3):227–242
Kumar S, Gupta S (2015) Crop diversification towards high-value crops in India: a state level empirical analysis. Agric Econ Res Rev 28(2):339–350
Thorat VA, Dhekale JS, Patil HK et al (2011) Determinants of rural-urban migration in Konkan region of Maharashtra. Agric Econ Res Rev 24:503–509
Turok I, McGranahan G (2013) Urbanization and economic growth: the arguments and evidence for Africa and Asia. Environ Urban 25(2):465–482
Udaykumar MS, Umesh KB, Srikantha Murthy PS (2020) Transaction costs in borrowing agricultural credit by farm households across rural-urban interface of Bengaluru. Curr J Appl Sci Technol 39(39):20–28

Chapter 5
Dairy Production in an Urbanising Environment: A Social-Ecological Approach to Understand the Case of Bengaluru

Marion Reichenbach, Ana Pinto, Sven König, Raghavendra Bhatta, and Eva Schlecht

Abstract As urban and rural worlds are increasingly pervaded by urban development, urban livestock husbandry is often considered as a phase-out model. Bengaluru is a fast-growing megacity in South India, where cows are a common sight everywhere. To understand the presence of cows throughout the rural–urban interface of Bengaluru, a baseline survey was carried out with 337 dairy producers. In terms of socio-economic profile of the dairy producers and dairy practices, the dairy sector was homogenous in the entire study region. Urban and some rural producers relied on public lands for pasture and fodder supply to deal with the constraint of land competition. Moreover, cows are part of a complex social-ecological system, providing multiple services to dairy producers, the consumers and the society in general.

Keywords Livestock · Dairy production systems · Semi-structured baseline survey · Productivity · Social-ecological services

M. Reichenbach (✉) · E. Schlecht
Animal Husbandry in the Tropics and Subtropics, University of Kassel, Witzenhausen and Georg-August-University Göttingen, 37213 Göttingen, Germany
e-mail: marion.reichenbach@uni-kassel.de

E. Schlecht
e-mail: eschlech@uni-kassel.de

A. Pinto · S. König
Animal Breeding and Genetics, Justus-Liebig-University Gießen, 35390 Gießen, Germany
e-mail: ana.pinto-garcia@agrar.uni-giessen.de

S. König
e-mail: sven.koenig@agrar.uni-giessen.de

R. Bhatta
ICAR—National Institute of Animal Nutrition and Physiology, Bengaluru 560030, India
e-mail: directornianp@gmail.com

E. Hoffmann et al. (eds.), *The Rural-Urban Interface*, The Urban Book Series,
https://doi.org/10.1007/978-3-030-79972-4_5

5.1 Introduction

As India's economy grows, its agroecosystems are pressured by drivers of change, among them is accelerated urbanisation. Thereby, urbanisation is a broad term, involving fast population growth at high density that pushes food systems towards production increase and intensification. Additionally, urbanisation involves urban sprawl over arable land used for pasture and forage cultivation, and fragmentation of the agricultural landscape, which complicates the provision of agroecosystem services.

Census data over the last decades show that Bengaluru's growth rate was among the highest in India (Verma et al. 2017); cows, however, are still part of its urban landscape, despite the obvious constraints cities represent for livestock (Butler 2012). Though cows are easy to spot in the city's streets, the dairy producers behind them and the dairy production systems along the city's rural–urban interface (RUI) have been little documented. For example, Roessler et al. (2016) highlighted different peri-urban livestock production systems and their dynamics of intensification or extensification in the West African cities of Tamale, Ghana and Ouagadougou, Burkina Faso. Daburon et al. (2017) reported different characteristics and adaptation mechanisms of peri-urban dairy producers that allowed them to take advantage of and deal with the opportunities and constraints represented by the urbanisation of Greater Cairo, Egypt. The following empirical data constitute a first step towards characterising Bengaluru's dairy production systems in relation to their spatial location. Broadened by anecdotal evidence, they point to a complex social-ecological system including dairy producers, the consumers and the cows.

5.2 Methodology

A two-step random selection process was used to survey a minimum of 300 dairy units across the northern and southern transect: 30 settlements were first selected along Bengaluru's RUI (Hoffmann et al. 2017). Within each settlement, 20% of all dairy units were randomly selected based on the latest vaccination list for Foot-and-Mouth disease.[1] To compensate for the low number of dairy producers in urban settlements, the selection threshold was increased to 30% in the more urban southern transect, and two urban settlements were purposefully added: Puttanahalli in the northern transect (2 surveys) and Shivaji Nagar, outside of both transects, next to the city centre (4 surveys). To assure potentially continuous, even though minimal, involvement in

[1] Mandatory vaccination campaign done every 6 months, indicating the name of the cattle owner and number of vaccinated cattle [personal communication by local veterinary officials].

Table 5.1 Main variables (mean ± standard deviation) characterising dairy production units (n = number) along the rural–urban interface of Bengaluru

Variable per settlement	Rural ($n = 14$)	Peri-urban ($n = 10$)	Urban ($n = 8$)
Avg. number of dairy units per settlement	55 ± 26^{a}	45 ± 37^{ab}	9 ± 7^{b}
Variable per dairy farm	Rural ($n = 197$)	Peri-urban ($n = 110$)	Urban ($n = 30$)
Dairy herd size (LDH[1], n)	2.8 ± 1.2^{a}	3.0 ± 1.1^{a}	4.8 ± 3.1^{b}
Milk production per cow (l/day)	7.5 ± 4.1^{a}	9.0 ± 3.9^{ab}	9.6 ± 5.2^{b}
Average price per litre of milk (INR[2])	23 ± 2^{a}	24 ± 1^{a}	25 ± 4^{b}
Dairy producers cultivating forage (%)	87	78	0
Dairy herds going to pasture (%)	71	80	80

[1]LDH = lactating (L) or dry cow (D), pregnant heifer (H)
[2]INR = Indian rupees; average price weighted according to the volume of milk sold to each marketing channel, without subsidy
Values within a row with different superscript letters significantly differ ($p < 0.05$)

milk marketing, only dairy producers with two or more dairy cattle (lactating (L) or dry cow (D), pregnant heifer[2] (H)) were considered.

The semi-structured baseline survey covered all topics related to dairy production: socio-economic characteristics of the household, dairy herd characteristics and management, inputs and outputs into the dairy unit and an open question about the motivations to produce milk. In total, 337 dairy producers were interviewed between mid-August and mid-November 2016.

5.3 Results

5.3.1 Similarities and Differences in Dairy Production Along the RUI

The number of dairy units per settlement was the first changing variable along the RUI: it decreased sharply in urban settlements despite the larger population, while peri-urban and rural settlements had similar numbers of dairy producers (Table 5.1).

Interestingly, most dairy producers had a similar socio-economic profile: a married male, in his fifties (average), head of a household of five members over three generations. Dairy units were family run and kept a similar number of cows with usually two or three household members participating in the dairy activity. In half of the

[2] Pregnant heifers were given similar care and feeding as cows within the herd. Dairy producers usually sold an older cow when pregnant heifers came close to calving, rather than to increase the size of their herd; this justified considering heifers as productive dairy cattle.

dairy units, at least one household member pursued an off-farm activity, but 61% of all dairy producers stated dairy production as their main source of income.

The average dairy herd comprised 3.0 ± 1.5 dairy cattle (LDH) plus 1.3 ± 1.2 female calves/weaners with larger herd sizes in urban settlements. Jersey and Holstein Friesian cattle constituted the largest share in all herds, either as purebred or first-generation crossbred.[3] The remaining cattle were crossbreeds of all types, and only 14% of all dairy producers owned some native cows (Hallikar breed).

Daily milk production per cow averaged 7.5 L in rural settlements but 9.6 L in urban ones. On average, 12% of the total daily production was kept for own consumption. In rural areas, extra milk was delivered to the settlement's dairy cooperative, associated to the Karnataka Milk Federation (KMF), where the price per litre was based on fat content and amended by a subsidy.[4] In urban areas, dairy producers marketed part of their extra milk for a higher price[5] directly to consumers or sold it to middlemen, and only leftover milk was delivered to a dairy cooperative.

Three quarters of all dairy producers relied at least partially on forage they cultivated themselves on own or rented land, or both. Maize and Napier grass were the main cultivated green forages, and finger millet (ragi) straw was offered during drier times of the year. If the dairy producer cultivated forages, the cows' manure was used as fertilizer. In addition to the conventional roughages, dairy producers collected green forage in exchange for farm work, wild grasses from public grounds, wild grasses from lakeshores, and, in the city, market wastes (fruits, vegetables) to supply their cows with low or no cost roughage. All dairy producers fed commercial concentrate feeds to their dairy cattle, and its purchase was unanimously stated as the main expense. If available, financial capital was first invested in the quality (genetics) of purchased cows and marginally in shed infrastructure, but not in a significant increase of herd size because of the reliance on solely family labour.

When asked which problem dairy producers faced most and what would affect their future involvement in dairy production, a common answer was "maintenance." Maintenance comprises all physical labour needed to feed the cattle (manual cultivation and harvest of forage, unmotorised forage transport), to take care of the cows (cleaning the shed, collecting and disposing manure) and to milk them (by hand). It was further considered difficult to provide enough forage to the cattle.

[3] Holstein Friesian heifers were usually inseminated with Jersey semen to ease their first calving, making first-generation exotic crossbreeds very common.

[4] In Indian rupees (INR): on average 23 INR (according to fat content) + 4 INR subsidy per liter = 0.36 euro at the time of the survey.

[5] On average 32 INR = 0.43 euro at the time of the survey.

5.4 Discussion

5.4.1 *A Social-Ecological Approach: The Consumer, the Producer and the Cow in the City*

Because almost one third of all Indians are vegetarian (Office of the Registrar General & Census Commissioner 2014), dairy products remain a major diet component despite the change in diet patterns usually observed with urbanisation (Chaps. 12–14), and Bengaluru's dairy producers have no problem selling their production. Raw fresh milk is considered even healthier than processed dairy products. Because consumers keep the tradition to boil it and process their own curd (yoghurt), ghee (butter) or buttermilk, consumer safety concerns for direct milk marketing are low. When diversifying farming systems, keeping livestock is a preferred choice, enhanced by the availability of credit or capital (Chaps. 4 and 6) and also reflected in the persistence or even increase of fodder cultivation around Bengaluru (Chaps. 2 and 3). To focus on milk as the sole contribution of dairy production to society is, however, of limited use for understanding the state of Bengaluru's dairy sector and the integration of cows in its urban landscape (Fig. 5.1).

First, the services provided by dairy production to the producer and the consumer also include other tangible outputs (manure as field fertilizer, cow urine as traditional medicine) and large intangible cultural and social outputs: as a holy animal, their presence is enjoyed and they are still part of many religious ceremonies, such as blessing of a new house, and their presence is enjoyed (Narayan 2018). Urban milk production contributes to city resilience through the preservation of a direct link between producers and consumers, employment and food security (Lerner and Eakin 2011; Deutsch et al. 2013; Cumming et al. 2014). Income generation though milk production contributes to the integration of dairy producers in the urban landscape, which allows them to continue their traditional economic activity, while enabling their children to benefit from better urban infrastructures (hospitals, schools).

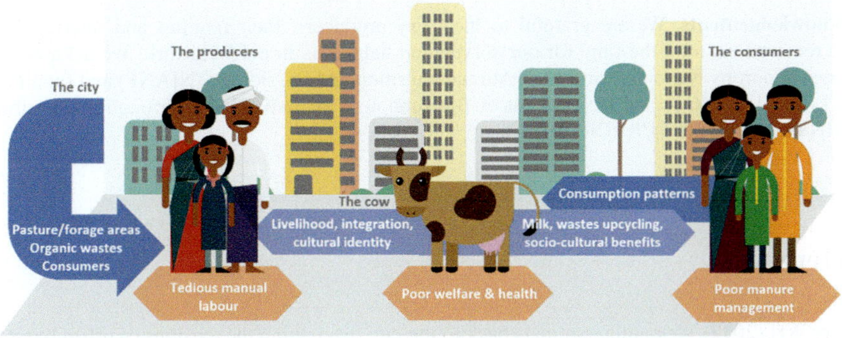

Fig. 5.1 Linkages (in blue) and constraints (in red) within the social-ecological system of Bengaluru's urban dairy production

Second, cows also provide interesting ecological services such as upcycling of low-value market wastes into highly valuable animal protein and manure but also disservices, for example, the risk of water pollution in urban settlements where manure management is poor.

At last, even though producers and consumers benefit from urban dairy production, the cows have to pay the price of impaired health, narrow and ill aerated shed space, inorganic wastes and plastic consumption, high risk of road accidents and a generally compromised well-being (Chap.10).

5.5 Conclusions

Overall, the increased competition for land that constrains forage production due to urbanisation is outweighed by increasing milk demand and the resourcefulness of urban dairy producers to find alternative forage sources or feedstuffs. Because of the bundle of services provided by cows, both producers and cows have their place within the urban landscape to the benefit of consumers. The city also benefits from cattle relying on and upcycling available organic resources ("wastes") and from their employment, income and food generating functions. As a megacity also represents a multitude of alternative economic opportunities, dairy producers in urban settlements may switch to new, better and economic activities. This puts into question the future of urban dairy production: as the new generation of city dwellers benefits from better education and job opportunities, the economic interest in dairy production will weaken, and even though small-scale urban dairy producers are flexible and resilient, they may not survive without institutional support. Therefore, Bengaluru's policy makers should support urban dairy production to more comprehensively harness its multiple services and economic potential, or focus on commercial (rural) production to match the urban demand for milk if urban dairy production is considered a phase-out model. Leaving the sector without a clear development strategy is a missed opportunity.

Acknowledgements We are grateful to the dairy producers, their families and intermediaries who received us, took the time for our survey and helped us in our fieldwork. We acknowledge strong support by National Institute of Animal Nutrition and Physiology (NIANP) and funding by DFG (SCHL 587/6-1) and Department of Biotechnology of India (DBT) through grant number BT/IN/German/DFG/14/BVCR/2016.

References

Butler WH (2012) Welcoming animals back to the city: navigating the tensions of urban livestock through municipal ordinances. J Agric Food Syst Commun Dev 2(2):193–215

Cumming GS, Buerkert A, Hoffmann EM et al (2014) Implications of agricultural transitions and urbanization for ecosystem services. Nature 515(7525):50–57

Daburon A, Alary V, Ali A et al (2017) Urban farms under pressure: Cairo's dairy producers, Egypt. In: Soulard CT, Perrin C, Valette E (eds) Toward sustainable relations between agriculture and the city. Springer, Cham, pp 73–88

Deutsch L, Dyball R, Steffen W (2013) Feeding cities: food security and ecosystem support in an urbanizing world. In: Elmqvist T, Fragkias M, Goodness J et al (eds) Urbanization, biodiversity and ecosystem services: challenges and opportunities. Springer, Dordrecht, pp 505–537

Hoffmann EM, Jose M, Nölke N et al (2017) Construction and use of a simple index of urbanisation in the rural–urban interface of Bangalore, India. Sustainability 9(11):2146–2157

Lerner AM, Eakin H (2011) An obsolete dichotomy? Rethinking the rural–urban interface in terms of food security and production in the global south. Geogr J 177(4):311–320

Narayan S (2018) The cows of Bangalore: and how I came to own one. Simon and Schuster, Noida

Office of the Registrar General & Census Commissioner, India (2014) Sample registration sys-tem, Baseline Survey 2014. Accessed on 15 Oct 2019. http://www.censusindia.gov.in/vital_statistics/BASELINE%20TABLES07062016.pdf

Roessler R, Mpouam S, Muchemwa T et al (2016) Emerging development pathways of urban livestock production in rapidly growing West Africa cities. Sustainability 8(11):1199–1224

Verma S, Chatterjee A, Mandal NR (2017) Analysing urban sprawl and shifting of urban growth centre of Bengaluru city, India using Shannon's Entropy method. J Settlements Spat Plann 8(2):89–98

Chapter 6
Capital Formation Across the Rural–Urban Interface of Bengaluru North: An Economic Analysis

H. V. Harish Kumar and B. V. Chinnappa Reddy

Abstract Pervading urbanisation creates uncertainty in prospects of agriculture across the rural–urban interface, which may affect capital formation for productive farm capital assets. To assess the capital investment behaviour of households north of Bengaluru and to identify factors influencing it, the present study used primary data at 2016–17 prices. Capital formation per hectare decreased from urban to peri-urban and rural areas (₹1.20 million, 0.67 million and 0.40 million, respectively). Investment on irrigation structure was highest throughout and formed half of the total farm investment in the rural area, owing to the dependence of agriculture on ground water. Another significant share of total capital was invested on labour saving farm machinery (18% in urban and peri-urban areas). Investment on livestock increased with urban influence and became a major livelihood source. The Cobb–Douglas type of production function revealed that land holding size significantly influenced the non-land capital stock throughout. The land sale status was non-significant in urban and peri-urban areas, whereas it was significant in the rural area, where most of the capital assets were generated out of land sale proceeds. The study concludes that urban influence increased investment on farm capital assets to cope with uncertainties in agriculture, namely acute labour shortage, increasing competition for water and increasing demand for livestock products.

Keywords Farm assets · Capital stock · Investment preferences · Livestock · Perennial crops · Cobb–Douglas type production function

6.1 Introduction

'Capital—the produced means of production' is an important factor for the growth of any sector. All tangible reproducible assets intended for use in future production

---

H. V. H. Kumar (✉)
ICAR-Indian Agricultural Statistics Research Institute, New Delhi 110012, India
e-mail: harishpal0083@gmail.com

B. V. C. Reddy
Agricultural Economics, University of Agricultural Sciences Bangalore, Bengaluru 560065, India
e-mail: bvchinnappareddy@gmail.com

E. Hoffmann et al. (eds.), *The Rural-Urban Interface*, The Urban Book Series,
https://doi.org/10.1007/978-3-030-79972-4_6

are part of capital formation as per the concept of United Nations (Hooley 1967). Capital formation in agriculture may include farm machinery and implements, land developments, establishment of new orchards and plantations, irrigation infrastructure and other constructions. Investment in farm capital stock in India comes from both public and private sectors. The farm household investment is largely directed towards irrigation development, establishment of perennial horticultural plantations and lately on farm machinery and equipment (Bisaliah 2010). Capital formation in the agricultural sector has a positive impact in augmenting agricultural production (Rao and Gulati 1994). Thus, a decrease in capital accumulation in this sector may, in turn, have serious implications for food security.

Rich literature exists on capital formation explaining its composition, trends, pattern, determinants and its impact on growth of farm output. However, not many studies focussed on capital formation in the context of urbanisation. India is transforming from a rural and agrarian-based country to an urban and industry-based country. For the first time in India, the urban population exceeds the rural in absolute terms, meanwhile the level of urbanisation has increased from 27.81% in 2001 to 31.16% in 2011 (Anonymous 2011). Urbanisation is also one of the important variables influencing capital formation in agriculture (Srivatsava 1986). Urban sprawl across space creates uncertainty in the prospects of agriculture, which may affect capital formation for productive farm capital assets. Hence, the nexus between urbanisation and capital formation in agriculture needs further attention for unfurling the pattern of capital accumulation as influenced by varying levels of urbanisation. This study has made an attempt to outline the influence of urbanisation on capital formation in one of the fast-growing cities of the world, Bengaluru. It adopted the concept of a spatial, rural–urban gradient, but extended the scope to the district level, considering the taluks Bengaluru North, Devanahalli, and Gudibanda as urban, peri-urban and rural areas, respectively.

6.2 Materials and Methods

This study was carried out across three study areas representing urban, peri-urban and rural scenarios which were identified based on their geographical location northward of Bengaluru. The taluk Bengaluru North (in the district Bengaluru Urban) was considered as urban area, Devanahalli (in Bengaluru Rural) as peri-urban and Gudibanda (in Chikkaballapur) as rural area. A multistage random sampling procedure was applied for selecting four study villages in each area: Bagaluru, Hosahalli, Srinivasapura and Singanayakanahalli in the urban area, Attibele, Gobbaradagunte, Venkatagirikote and Chikkamarali in the peri-urban, Chendur, Bhandarlahalli, Thatahalli and Doddakurubarahalli in the rural area. Fifteen respondents from each village were randomly selected for the study. Thus, primary data on farm capital stock were collected from 60 sample farm households in each area within the rural–urban gradient through personal interview using a pre-tested structured schedule developed for the said purpose. The farm capital stock was generated over the period from 1993–94 to 2016–17. Hence, the investment on each capital asset was valued at 2016–17 prices (compounding process) by using the wholesale price index.

Capital formation in agriculture is influenced by various socio-economic variables. In the household interviews, six highly influential factors were identified to be taken up in the production function. Among various forms of production functions, the Cobb–Douglas (C–D) type of production function was best fitted for the study data. Variables like education and land sold status were of the nominal scale; hence, they were introduced using dummy. The formula of the C–D production function used in this study is as follows (Eq. 6.1):

$$\ln Y = \ln a + b_1 \ln X_1 + b_2 \ln X_2 + b_3 \ln X_3 + b_4 \ln X_4 + b_5 D_1 + b_6 D_2 + u_i \tag{6.1}$$

where

Y	Non-land farm capital stock (in million ₹)
X_1	Land holding size (in ha)
X_2	Number of farm youth (in numbers)
X_3	Net farm income (million ₹)
X_4	Loan amount borrowed (million ₹)
D_1	Dummy for education status (Illiterate = 0, otherwise = 1)
D_2	Dummy for land sold status (No = 0, otherwise = 1)
a	Constant
b_1–b_6	Coefficients of respective variables and u_i = Error term.

6.3 Results

6.3.1 *Pattern of Farm Capital Formation*

The extent of capital formation in agriculture was expressed on a per hectare basis (Table 6.1). It decreased along with the presumed urban influence, i.e. it was highest in the urban (₹ 1.206 million) followed by the peri-urban area (₹ 0.670 million) and was least in the rural area (₹ 0.398 million).

In urban and rural areas, investment in irrigation infrastructure (36.2% and 50.3%) constituted the major part of farm capital stock, followed by farm machinery (17.9 and 13.8%). In the peri-urban area farm capital stock mainly constituted investment in farm buildings (30.5%), followed by irrigation infrastructure (22.3%). Per hectare investment in livestock was higher in urban (₹ 0.142 million) than peri-urban (₹ 0.096 million) and rural (₹ 0.047 million) parts of the gradient.

Table 6.1 Investment in farm capital assets (₹ million per hectare) in Bengaluru, India

Sl. No	Farm capital assets	Urban	Peri-Urban	Rural
1	Irrigation infrastructure	0.438 (36.32)	0.149 (22.28)	0.200 (50.26)
2	Farm machinery	0.215 (17.86)	0.118 (17.63)	0.055 (13.78)
3	Farm buildings	0.156 (12.91)	0.204 (30.48)	0.041 (10.20)
4	Perennial crops	0.154 (12.74)	0.098 (14.57)	0.053 (13.27)
5	Livestock	0.142 (11.79)	0.096 (14.32)	0.047 (11.73)
6	Purchase of land/improvement on land	0.101 (8.38)	0.005 (0.73)	0.003 (0.77)
Total		1.206	0.670	0.398

Note values in parenthesis are percentage share of farm capital asset to total farm capital stock

6.3.2 Factors Influencing Capital Formation in Agriculture

To analyse the influence of socio-economic variables on capital formation in agriculture in the context of urbanisation, non-land farm capital stock was regressed against six identified variables using a C–D type production function (Table 6.2).

In the urban area, the independent variables under consideration explained 62% of variation in non-land farm capital stock, and the model had good fit and was statistically significant based on the F value at one per cent level. The variable "net farm income" was considered here as exogenous, as it will affect a farmer's ability to invest and, thus, the farm's capital stock, and as such, it was non-significant throughout. The variables land holding size, loan amount borrowed and numbers of youth were significantly influencing the non-land farm capital stock.

In the peri-urban area, land holding size was the only variable influencing non-land capital stock significantly with the elasticity of 1.33%. The co-efficient of multiple determination (R^2) was relatively low (0.49), but the model was significant ($P < 0.05$). In the rural region, along with land holding size, land sold status was also significantly influencing non-land farm capital stock. The variables included in this model explained 64% of variation in the dependent variable, and the model was significant. In general, land holding size was significantly influencing the non-land capital stock across all three areas of the gradient.

Table 6.2 Regression results of Cobb–Douglas (C–D) production function

Sl. No	Parameters	Urban	Peri-Urban	Rural
1	Intercept	0.66[a] (0.54)	−2.45[a] (-0.78)	−0.83[a] (−1.22)
2	Land holding size [hectares]	1.17*** (4.38)	1.33*** (4.09)	1.25*** (3.03)
3	Number of youth with household [numbers]	0.83** (2.26)	0.24[a] (0.48)	−0.64[a] (−1.78)
4	Net farm income [million ₹]	0.0[a] (0.14)	0.24[a] (0.85)	−0.02[a] (−0.24)
5	Loan amount borrowed [million ₹]	0.08** (2.08)	0.00[a] (−0.03)	0.03[a] (1.15)
6	Education of respondent[Literate = 1, otherwise = 0]	−0.45[a] (-0.93)	−0.36[a] (−0.77)	0.13[a] (0.38)
7	Land sold [yes = 1, otherwise = 0]	−0.45[a] (−0.94)	−0.17[a] (−0.33)	0.89** (2.43)
	R^2	0.62*** [6.26]	0.49** [4.15]	0.64*** [7.48]

Note $^{***}p < 0.01$, $^{**}p < 0.05$
[a]*NS* non-significant
t values in () and *F* values in []

6.4 Discussion

Per hectare capital formation was highest in the urban area followed by peri-urban and rural areas due to intensive cultivation practice with a higher share of irrigation-based farming. Hence, the process of urbanisation had a positive impact on capital formation (Srivatsava 1986). In the entire study region, there was no other irrigation source except for ground water. The ground water table has been declining continuously over the last two decades, especially in intensely urbanised areas of Bengaluru (Ramachandra et al. 2016). Hence, investment in irrigation infrastructure forms a major share in total farm capital stock as massive investment is required for the drilling of bore wells and installation of the unit. Labour shortage is another major issue in the study region, more specifically in the urban and peri-urban areas; hence, considerable investment was made on farm machinery and equipment. A similar pattern of investment was reported in Hoskote taluk of Bengaluru rural district (Parameswarappa 1988). Farm buildings including cattle sheds attracted a sizable portion of investment especially in peri-urban villages. The study area is also known for perennial crops like grapes (*Vitis vinifera* L.), guava (*Psidium guajava* L.) and mango (*Mangifera indica* L.); hence, relatively high investment is made on such orchards all along the gradient (Chap. 4). Livestock is a key livelihood of the households in the study area especially in urban and peri-urban villages (Chap. 5; Shobha et al. 2020). Thus, per hectare investment was found to be higher than in the rural area. In the process of urbanisation in and around Bengaluru, analogous patterns of investment on farm capital assets were reported for highly (Kolar taluk) and less progressive areas (Magadi taluk; Venkataramana 2010).

Land holding size significantly influenced the non-land capital stock all along the gradient and one per cent increase in land holding size would lead to increase in non-land farm capital stock by 1.17%, 1.33% and 1.25% in urban, peri-urban and rural gradient, respectively. So in general, the size of land holding had a bearing on capital formation in agriculture (Srivatsava 1986; Bisaliah et al. 2010; Venkataramana 2010). Net farm income had no significant influence on capital stock when considered as exogenous variable. On the other hand, a farm's investments and capital stock may also have an effect on net farm income, so that causality might run in both directions, and net farm income could also be considered endogenous rather than exogenous. To deal with this, however, would require a more complex model, which is beyond scope of the present paper. Under the available land holding, net sown area is decreasing in the areas under high urban influence due to developmental pressures on farm land for non-agricultural uses (Harishkumar and Reddy 2017a; Te Lintelo et al. 2001; Rimal 2013; Rani 2014 and Kavitha et al. 2015). Hence, decreasing net sown area may affect the capital formation in agriculture in these areas in future even though per hectare farm capital stock is higher in urban and peri-urban gradients. The land sale status was significantly influencing non-land farm capital stock only in the rural region, where most of the farm capital assets were generated out of land sale proceeds; whereas in urban areas, the land sale proceeds were mainly used for wasteful consumption and investment in unproductive assets (Harishkumar and Reddy 2017b). Hence, farm capital accumulation out of farm land sale proceeds needs to be encouraged. Private capital formation is also positively influenced by the institutional credit as it has inducement effect (Bisaliah and Dev 2010; Chand and Kumar 2004; Chap. 4).

6.5 Conclusions

The dimension of capital formation in agriculture in the context of pervading urbanisation needs attention. The data revealed that urban influences have brought about more investment on farm capital assets to cope with uncertainties in agriculture. Though per hectare farm capital stock was relatively high in more urban influenced areas, the concern of decreasing net sown area may pose a serious threat on future gross capital formation in agriculture. Hence, there is a need to sensitise the farm households in this regard in areas under strong urban influence.

Further insights to the investment behaviour of farm households may be gained if the direction of causality of variables such as net farm income and farm's capital stock was analysed by more complex models that accounted for potential positive or negative feedback loops. This is a topic for future research.

Acknowledgements We acknowledges DST-Inspire, Ministry of Science and Technology, GoI for awarding the Inspire Fellowship to carry out this work under the Ph.D. research programme.

References

Anonymous (2011) Census of India. Government of India New Delhi

Bisaliah S (2010) Capital formation, agriculture growth and poverty. Conceptual and empirical constructs. Consultancy Report, Food and Agriculture Organization Rome

Bisaliah S, Dev SM (2010) Private capital formation in Indian agriculture: an analysis of farm level data. Consultancy Report, Food and Agriculture Organization Rome

Chand R, Kumar P (2004) Determinants of capital formation and agriculture growth: Some new explorations. Econ Pol Weekly 39(52):5611–5616

Harishkumar HV, Reddy (2017a) Farm land values and portfolio management in rural—urban continuum of Bengaluru North Mysore. J Agric Sci 51(2):208–215

Harishkumar HV, Reddy BVC (2017b) Impact of urbanization on land use pattern of rural-urban gradient of Bengaluru North: an economic analysis Econ Affairs 62(2):303–312

Hooley RW (1967) The measurement of capital formation in underdeveloped countries. Rev Econ Stat 49:199–208

Kavitha A, Somashekar RK, Nagaraja BC (2015) Urban expansion and loss of agriculture land—a case of Bengaluru city. Int J Geomatics Geo-Sci 5(3):492–498

Parameshwarappa KJ (1988) Pattern of income, investment and employment in irrigated farms of Hoskote taluk, Bengaluru Rural district. Dissertation, University of Agricultural Sciences Bangalore

Ramachandra TV, Vinay S, Mahapatra DM et al (2016) Water situation in Bengaluru. Technical Report, Indian Institute of Science Bengaluru

Rani S (2014) Monitoring land use/land cover response to urban growth of the city of Jalandhar using remote sensing data. Int J Adv Res 2(6):1122–1129

Rao CHH, Gulati A (1994) Indian agriculture: emerging perspectives and policy issues. Econ Pol Weekly 29(53):158–169

Rimal B (2013) Urbanization and the decline of agricultural land in Pokhara sub-metropolitan city. Nepal J Agric Sci 5(1):54–65

Shobha KA, Venkataramana MN, Roy A et al (2020) Economic analysis of dairy production among small and medium scale farmers in Karnataka: a case study of Bengaluru district. Ind J Ani Sci 90(5):798–803

Srivatsava GC (1986) Impact of urbanization on capital formation in agriculture—a case study at Ranchi (Bihar). Agric Situation India 40(2):689–697

Te Lintelo D, Marshall F, Bhupal DS (2001) Peri-urban agriculture in Delhi, India. Food Nutrition Agric 29:4–13

Venkataramana MN (2010) Capital formation in agriculture-An economic study in eastern dry zone of Karnataka. Dissertation, University of Agricultural Sciences Bangalore

Part III
Resource Use, Efficiencies and Tradeoffs in Crop and Livestock Production

Chapter 7
Crop Production Under Urbanisation: An Experimental Approach to Understand and Model Agricultural Intensification

Andreas Buerkert, Ellen Hoffmann, Renuka Suddapuli Hewage, Sven Goenster-Jordan, Suman Kumar Sourav, Andrea Mock, Prem José Vazhacharickal, C. T. Subbarayappa, Mudalagiriyappa, D. C. Hanumanthappa, Stephan Peth, and Michael Wachendorf

Abstract Rural–urban transformation has major implications on agricultural land use. This is also the case in the southern Indian city of Bengaluru, where farmers shift from low intensity subsistence agriculture under rainfed conditions to irrigated, market-oriented production of crops and vegetables. As little is known about the effects of this intensification on water use, nutrient leaching, losses of carbon and nitrogen, and soil quality, a long-term experiment was established under well-defined

---

A. Buerkert (✉) · E. Hoffmann · R. Suddapuli Hewage · S. Goenster-Jordan · S. K. Sourav · A. Mock · P. J. Vazhacharickal
Organic Plant Production and Agroecosystems Research in the Tropics and Subtropics (OPATS), University of Kassel, Steinstrasse 19, 37213 Witzenhausen, Germany
e-mail: buerkert@uni-kassel.de

E. Hoffmann
e-mail: ellen.hoffmann@uni-kassel.de

S. Goenster-Jordan
e-mail: goenster@uni-kassel.de

S. K. Sourav
e-mail: sumanks@uni-kassel.de

A. Mock
e-mail: a.mock@uni-kassel.de

P. J. Vazhacharickal
e-mail: premjosev@gmail.com

M. Wachendorf
Grassland Science and Renewable Plant Resources, University of Kassel, Steinstrasse 19, 37213 Witzenhausen, Germany
e-mail: mwach@uni-kassel.de

C. T. Subbarayappa
Soil Science and Agricultural Chemistry, University of Agricultural Sciences Bangalore (UASB), GKVK Campus, Bengaluru 560065, India
e-mail: ctsubbarayappa@gmail.com

E. Hoffmann et al. (eds.), *The Rural-Urban Interface*, The Urban Book Series,
https://doi.org/10.1007/978-3-030-79972-4_7

on-station conditions to generate a typical intensity gradient in an in situ laboratory of change. Measurements of key agronomic, soil-related, and meteorological parameters at high temporal and spatial resolution allow to assess externalities and efficiencies of resource use and to predict long-term consequences of intensification on agricultural sustainability. The two long-term rotation experiments established under rainfed and irrigated conditions also allow to collect and calibrate ground-based multi- and hyperspectral crop reflectance data needed for upscaling to high resolution satellite images that cover a North–South research transect across the rural–urban interface of Bengaluru.

Keywords Agricultural production intensity · Crop rotation · Long-term cropping experiments · Nutrient fluxes · Reference data · Rural–urban transformation · Scenario analysis

7.1 Introduction

The sprawling development of Indian cities competes with agriculture for (peri-) urban land, water resources, and labor. This has led to major changes of the cropping pattern in the vicinity of urban areas and farmers focusing on high-priced horticultural or local specialty crops such as finger millet (*Eleusine coracana* Gaertn.) and grape (*Vitis vinifera* L.; see also Chaps. 2, 3, 19). These systems often have a lower water demand per unit of market value, but crops are typically cultivated in intensive rotations year-round or in drip irrigated settings (Jain et al. 2019). While it is well known that farmers continuously adapt their management systems in response to new constraints and urban demands (Swain and Teufel 2017), major knowledge gaps exist about the effects of year-round crop growth on matter fluxes and negative externalities, particularly the volatilization of carbon (C) and nitrogen (N), and the leaching of N, phosphorus (P), potassium (K), and sulfur (S).

During the last decade, numerous studies provided evidence of large C and N emissions from intensive (sub-)tropical agricultural systems. These emissions reflect rapid mineralization following continuous cultivation and regular annual inputs of organic and mineral fertilizers (Diogo et al. 2010; Predotova et al. 2010; Lompo et al. 2012) under year-round high temperature conditions. In most cases, however, it

Mudalagiriyappa
AICRP Rainfed Agriculture, University of Agricultural Sciences Bangalore (UASB), GKVK Campus, Bengaluru 560065, India

D. C. Hanumanthappa
AICRP Agroforestry, University of Agricultural Sciences Bangalore (UASB), GKVK Campus, Bengaluru 560065, India
e-mail: dhdeeta@gmail.com

S. Peth
Soil Science, University of Kassel, Nordbahnhofstr. 1, 37213 Witzenhausen, Germany
e-mail: peth@ifbk.uni-hannover.de

remained unclear what proportion of such emissions was due to the turnover of the added substrate and how much to mobilization from different soil fractions of C and N. Moreover, little systematic information is available on the effects of irrigation on turnover processes, matter balances, and nutrient use efficiency in (sub-)tropical agroecosystems of different intensity such as the ones widely practiced in monsoon-driven southern India.

To fill some of these knowledge gaps, we designed a set of two factorial experiments on the premises of the University of Agricultural Sciences Bangalore (UASB). They comprise six major local crops in a typical rotation with a three-year cycle. To cover the range of management intensities found in the rural–urban interface of Bengaluru, we apply three levels of N-fertilization with and without (rainfed) irrigation. Tight measurements of crop phenology and yield, nutrient input and outputs including leaching and gaseous losses, a wide range of soil parameters including soil moisture and temperature, and of agro-meteorological conditions at 10-min intervals, provide a unique research infrastructure to mimic scenarios of agricultural intensifications and comparatively assess the externalities of typical farming practices along the rural–urban gradient of Bengaluru.

7.2 Materials and Methods

7.2.1 Setup of the Field Experiments

Since May 2016, both experimental sites are established at the GKVK Campus, UASB, on typical deeply weathered Kandic Paleustalfs/Dystric Nitisols (Chap. 8). Located ca. 920 m a.s.l., the Indian summer monsoon season lasts from July to October with a mean annual rainfall of 873 mm (±SD 197 mm). The mean temperature is 26.7 °C in winter and 34.0 °C in summer (Murugan et al. 2019; Brinkmann et al. 2020). The two experimental sites are 500 m apart (see Chap. 8). Following local practice, crops are cultivated in the rainy season, *Kharif*, lasting from July/August to November/December, and the dry season, *Rabi*, from February to June. During *Kharif*, the same cereals and pulses are cultivated at both sites, with irrigation (IE) and without irrigation (RE), and during *Rabi*, vegetables are grown in the IE, while the RE is left fallow.

As shown in the introductory chapter (Chap. 1), finger millet/ragi (*Eleusine coracana* Gaertn.) and maize (*Zea mays* L.) are the most widely cultivated cereals in the region; lablab/avare (*Lablab purpureus* L. Sweet) is among the most abundant pulses, and tomato (*Solanum lycopersicum* L.), cabbage (*Brassica oleracea* L.), and eggplant (*Solanum melongena* L.) are popular vegetables produced by local farmers for the attractive urban markets. These species were, therefore, chosen for the experiment.

The factorial experimental design comprises three crop species per season and three N-fertilizer levels (low, medium, and high) at four replications. Thus in each of the two experiments, there are 12 main experiment plots (12 m × 18 m), each divided into three sub-plots (12 m × 6 m), yielding a total of 36 sub-plots per experiment with a specific combination of crop-by-N level (Figs. 7.1 and 7.2). The distribution

Fig. 7.1 Irrigated experiment during the *Rabi* season (above) and rainfed experiment (left below) and irrigated experiment (right below) during the *Kharif* season at GKVK Campus, UAS Bangalore, India

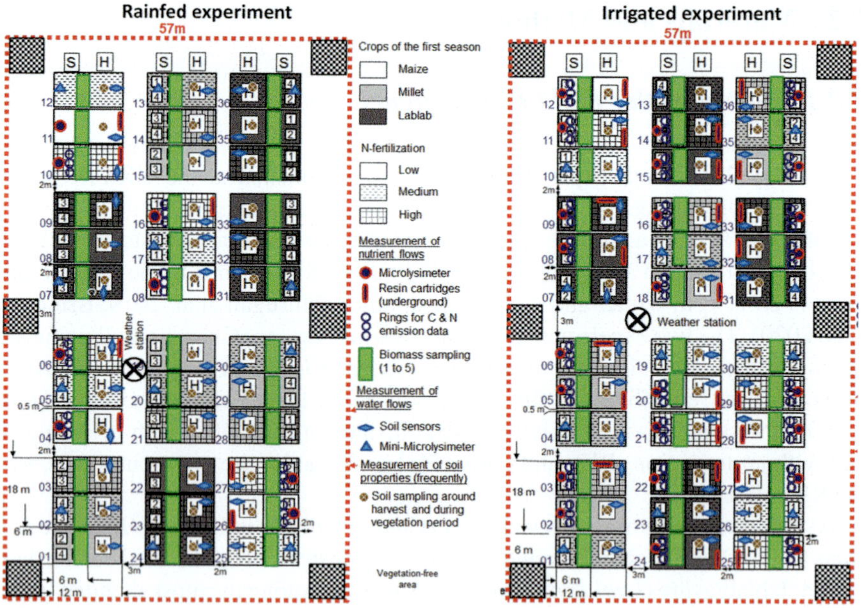

Fig. 7.2 Experimental plot layout of the two on-station experiments at GKVK Campus, UAS Bangalore, India (crop allocation in the 2016 *Kharif* season). Small numerical inserts refer to repeated destructive samplings in the S-sub-plot. Different shades of grey represent the crop species, different hatch-patterns the N-fertilization levels

Table 7.1 Crop rotation applied in the field experiments at GKVK Campus, UAS Bangalore, India. Typical succession on a plot that started the rotation with maize in Year 1 (2016). Plots that started with another crop followed the same sequence, such that after three years, all crops were cultivated once on each plot

		Rainfed experiment	Irrigated experiment
Year 1	*Kharif*	Maize	Maize
	Rabi	(fallow)	Cabbage
Year 2	*Kharif*	Millet	Millet
	Rabi	(fallow)	Eggplant
Year 3	*Kharif*	Lablab	Lablab
	Rabi	(fallow)	Tomato

of crops (in the main plots) and different N-fertilizer levels (in the sub-plots) were randomized at the beginning of the experiment. In successive seasons, crops are rotated (Table 7.1), while the N-fertilizer status in the sub-plots remain fixed for the entire duration of the experiment. The actual dosage of N-fertilizer, however, depends on the demand of the respective crop varieties: *low* is zero N for all crops, *medium* corresponds to the recommended crop-specific dosage by the local extension service, and *high* is twice the recommendation. The latter is sometimes applied by farmers eager to obtain highest yields (Table 7.2). Each sub-plot is further divided into two parts (6 m × 6 m), so that destructive sampling of plants or soil is confined to one half (called S sub-plot), while the other half (H sub-plot) is left untouched until final harvesting. To capture within-plot variation, a grid was laid over the H plot (25 cells with $(1.2\ m)^2$ per grid cell) for the final harvest of the *Kharif* crops. This paper presents data at the plot level only.

7.2.2 Determination of Biomass and Soil Nutrients

For each crop, fresh matter (FM) is determined in the field at final harvest. Dry matter (DM) of generative and vegetative parts is determined by representative samples dried at 100 °C to constant weight. The generative parts are usually the marketable yield, here understood as maize cobs, millet ears, lablab pods, tomato fruits, and eggplant fruits, while the vegetative fraction comprises stems and leaves. Accordingly (though not physiologically correct), the marketable cabbage head is recorded as the 'generative' fraction, and the outer leaves, which are discarded after harvest, as part of the vegetative fraction. Water content did not differ significantly between N-application levels, but clearly between plant parts (vegetative > generative) and slightly between places (RE < IE).

Samples of the topsoil (0–10 cm) were taken in all plots before the start of the experiment in 2016, and in 2019 after a full crop rotation cycle was completed. For soil analysis, standard methods are used (VDLUFA 2020): pH is determined in 1:2.5 water extracts, total C and N by a VarioMax CN analyzer (Elementar Analysensysteme GmbH, Hanau, Germany). Mineral N (NO_3-N, NH_4-N) and PO_4-P

Table 7.2 Summary of crop variety specifications and field management practices

Year	Season	Crop	Varieties	Duration (days)	Spacing (cm)	Field	Sowing date	Harvesting date	Mineral fertilizer application/season				
									N			P	K
									L	M	H		
2016	*Kharif*	Maize	Nithyashree	110–120	30 x 60	IE	19.07.2016	01.12.2016	50	100	150	75	50
						RE	19.07.2016	01.12.2016	50	75	100	50	37.5
		Finger millet	GPU 28	110–115	15 × 30	IE	19.07.2016	01.12.2016	50	75	100	50	50
						RE	19.07.2016	22.11.2016	25	37.5	50	40	37.5
		Lablab	HA4	100–105	15 × 30	IE	19.07.2016	25.11.2016	10	15	25	10	10
						RE	19.07.2016	17.10.2016	10	15	25	10	10
2017	*Rabi*	Egg plant	Ankur	55–60*	60 × 120	IE	22.02.2017	13.06.2017	25	50	75	40	20
		Tomato	NS 501	Early (65)**	60 × 120		22.02.2017	06.06.2017	23	46	69	40	24
		Cabbage	Unnati	80–85**	60 × 60		22.02.2017	15.06.2017	30	60	90	40	50
2017	*Kharif*	Maize	Nithyashree	110–120	30 x 60	IE	15.07.2017	11.11.2017	50	100	150	75	50
						RE	14.07.2017	09/10.11.2017	50	75	100	50	37.5
		Finger millet	MR 6	120–125	15 × 30	IE	15.07.2017	21/22.11.2017	50	75	100	50	50
						RE	14.07.2017	17/18.11.2017	25	37.5	50	40	37.5
		Lablab	HA 4	100–105	15 × 30	IE	15.07.2017	14/15.10.2017	10	15	25	10	10
						RE	14.07.2017	16.10.2017	10	15	25	10	10
2018	*Rabi*	Egg plant	Lalith	45–50*	60 × 120	IE	17/18.02.2018	11.06.2018	0	25	50 + 25[1]	40	20

(continued)

Table 7.2 (continued)

Year	Season	Crop	Varieties	Duration (days)	Spacing (cm)	Field	Sowing date	Harvesting date	Mineral fertilizer application/season				
									N			P	K
									L	M	H		
		Tomato	Arkha Rakshak	81**	60 × 120		17/18.02.2018	30.05.2018	0	23	46 + 23[1]	40	24
		Cabbage	Unnati	80–85**	60 × 60		17/18.02.2018	04.06.2018	0	30	60 + 30[1]	40	50
2018	*Kharif*	Maize	Hema	100–120	30 x 60	IE	07.08.2018	01.12.2018	0	50	150	75	50
						RE	13.08.2018	09.12.2018	0	50	150	50	37.5
		Finger millet	ML 365	110–115	15 × 30	IE	07.08.2018	12.12.2018	0	25	50	50	50
						RE	13.08.2018	01.12.2018	0	25	50	40	37.5
		Lablab	HA 3	95–100	15 × 30	IE	07.08.2018	15.11.2018	0	12.5	25	10	10
						RE	13.08.2018	15.11.2018	0	12.5	25	10	10
2019	*Rabi*	Egg plant	Lalith	45–50*	60 × 120	IE	18/19.02.2019	10.06.2019	0	25	50 + 25[1]	40	20
		Tomato	Arkha Rakshak	81**	60 × 120		18/19.02.2019	14.06.2019	0	23	46 + 23[1]	40	24
		Cabbage	Unnati	80–85**	60 × 60		18/19.02.2019	14.06.2019	0	30	60 + 30[1]	40	50

*days to first fruit picking, ** days to maturity
IE Irrigated experiment, *RE* rainfed experiment, *L* low N, *M* medium N, *H* high N, 1 only in H level received a 3rd supplementary N application 4 weeks after 2nd fertilizer application

are measured photometrically. Phosphorus is detected by colorimetry (Hitachi U-2000 spectrophotometer, Hitachi Ltd. Corp., Tokyo, Japan) according to the vanado-molybdate method (Gericke and Kurmies 1952), potassium (K) concentration is determined by flame photometry (Instrument Laboratory 543, Bedford, MA, USA) and SO_4-S after double lactate extraction (VDLUFA 2020).

7.2.3 Instrumentation for Monitoring of Nutrient Flows

The plots are equipped with sensors to monitor soil water content and nutrient flows, either continuously (SoilNet sensors, see Chap. 8) or cumulatively over defined periods of time with anion/cation resin cartridges (Friedel et al. 2000; Willich and Buerkert 2016) and lysimeters (Werner et al. 2019; Fig. 7.3). A closed chamber system consisting of a Teflon® coated PVC cuvette connected to a photo-acoustic infrared multi-gas analyzer (INNOVA 1312–5, LumaSense Technologies A/S, Ballerup, Denmark) and a portable spectroscope (Los Gatos Research Inc., Mountain View, CA, USA) is used to determine CO_2, N_2O, and NH_3 emissions from the soil surface of the plots throughout the seasons. Low-level aerial photography (Möckel et al. 2018), hyperspectral reflectance measurements (Astor et al. 2020, Chap. 9), and 3D terrestrial laser scanning (LiDAR; Jayakumari et al. 2021) are employed to calibrate reflectance measurements against upscaled spectral data from WordView 3 images taken in the *Kharif* season of 2017 and 2020.

The destructive and non-destructive, high frequency/high resolution data on crop yields and matter fluxes generated in both field experiments provide a unique inter-disciplinary opportunity to compare and model the effects of different management intensities. This will pave the way to upscale results from the local field scale to regional levels along the rural–urban gradient of Bengaluru. In this paper, we only present a first, limited set of agronomic results, which are complemented by

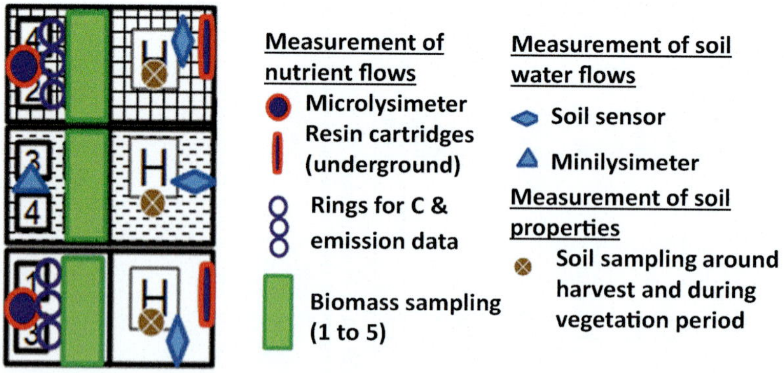

Fig. 7.3 Details of instrumental measurement setup in the rainfed and irrigated on-station experiments at GKVK Campus, UAS Bangalore, India

further data and comprehensive analyses. We also discuss the potential of the above-mentioned measurements for understanding the implications of the agricultural transitions observed in the rural–urban interface.

7.3 Results

7.3.1 *Biomass Yield*

Crop yields (Figs. 7.4 and 7.5) varied markedly between years, whereby in 2017, they were exceptionally high and 2018 exceptionally low reflecting major differences in monsoonal rainfall. Effects of N-fertilization and irrigation built up from the second year onwards; on average, yield under irrigation was almost twice as high as under rainfed conditions. Also initial analysis of yield variation within plots indicated its decline with irrigation and N-application over time.

7.3.2 *Soil Chemical Properties*

Since the onset of the experiments average soil pH was with 5.1 substantially higher in the irrigated field experiment than with 4.4 in the rainfed one, reflecting differences in field history (continued cropping in the rainfed field *versus* many years of eucalyptus forest in the irrigated one). Over the three-year cultivation period, this difference increased across N-fertilization to 6.4 *versus* 4.2, respectively.

Similarly, in 2016, soil C and N content were almost twice as high in the irrigated field and did not change much over the cultivation period (Fig. 7.6).

Topsoil (0–10 cm) N and K were higher in the irrigated field, whereas plant available Bray-P was higher in the rainfed field (Fig. 7.7). In 2019, despite substantial application as KCl, K was near detection limit at both locations.

7.4 Discussion and Conclusions

The high density of collected data in the controlled on-station experiments allows to determine the effects of typical differences in management intensity on soil quality parameters, yields, and horizontal (nutrient inputs *versus* harvested outputs) and vertical (leaching *versus* gaseous C and N emissions) matter fluxes for crops grown under conditions typical for farmers along the rural–urban gradient of Bengaluru (scenario analysis). The data also permit to test our initial hypothesis that yield variability declines over time and with increasing management intensity. This effect will likely be crop specific. The surprisingly high and stable crop yields in zero N

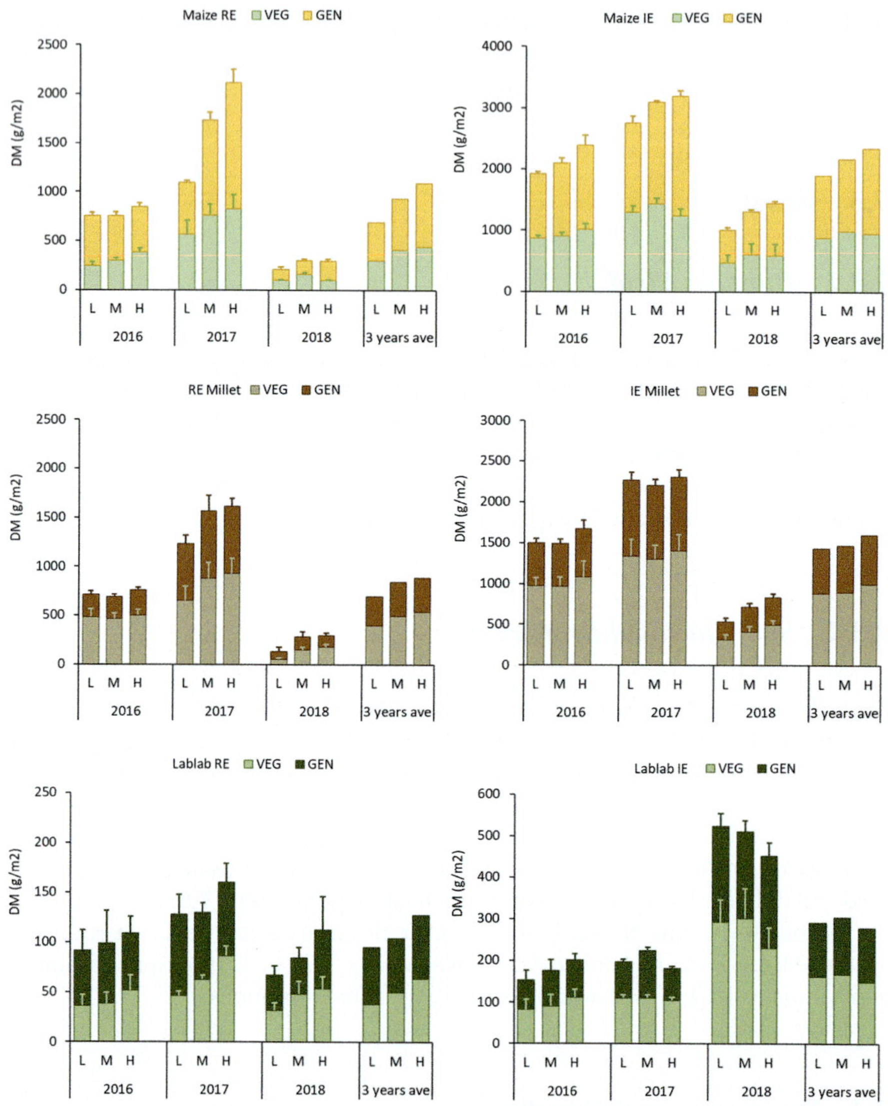

Fig. 7.4 Dry matter yields separated for vegetative (VEG) and generative (GEN) parts of maize, millet, and lablab under low (L), medium (M), and high (H) N-fertilization in the rainfed (left column) and irrigated (right column) on-station experiments at GKVK Campus, UAS Bangalore, India. Vertical bars show one standard error of the mean ($n = 4$)

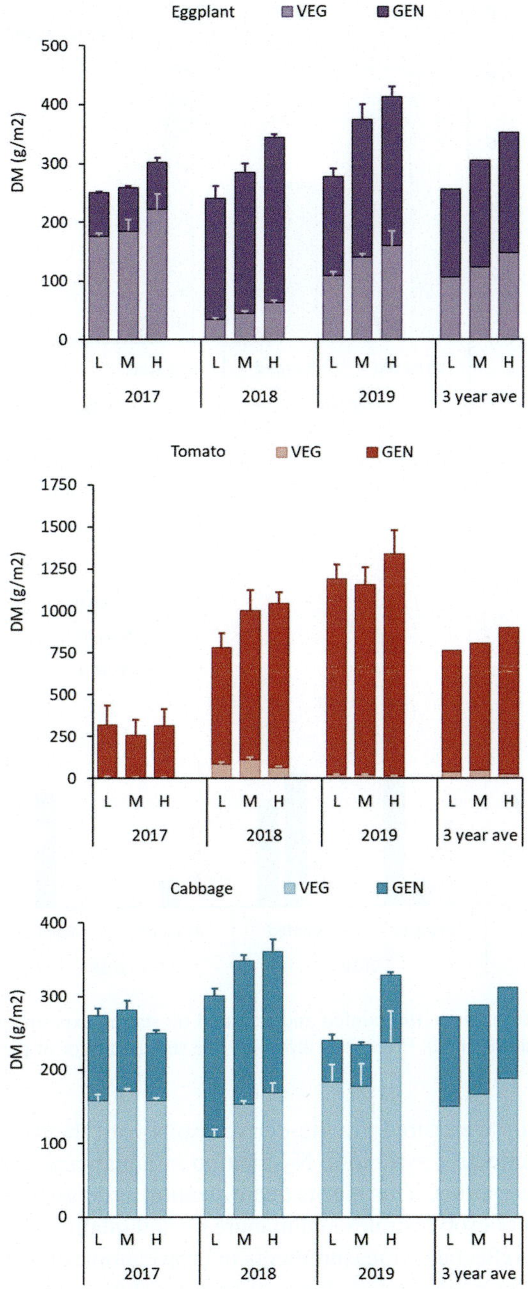

Fig. 7.5 Dry matter yields separated for vegetative (VEG) and generative (GEN) parts of eggplant, tomato, and cabbage under low (L), medium (M), and high (H) N-fertilization in the irrigated on-station experiments at GKVK Campus, UAS Bangalore, India. Vertical bars show one standard error of the mean ($n = 4$)

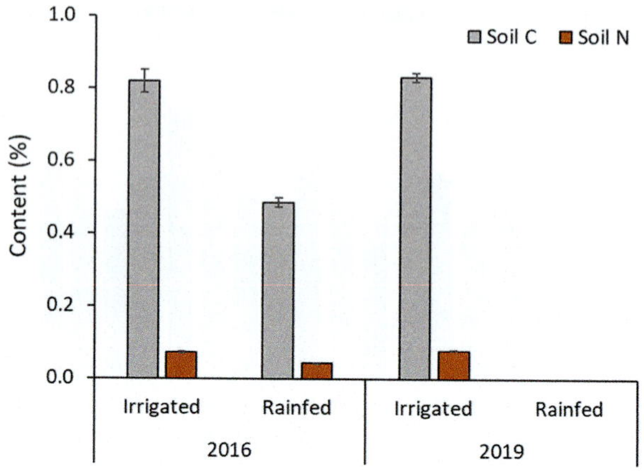

Fig. 7.6 Soil carbon and total nitrogen content in the rainfed and irrigated on-station experiments at GKVK Campus, UAS Bangalore, India. Error bars show +/- one standard error of the mean (n = 36)

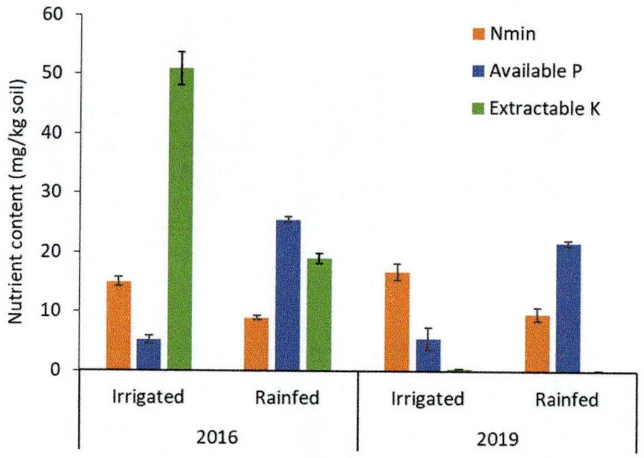

Fig. 7.7 Soil nutrient status in the rainfed and irrigated on-station experimental fields at GKVK Campus, UAS Bangalore, India. Error bars show +/- one standard error of the mean (n = 36)

plots on the heavily weathered granite-derived soils with their low organic carbon contents point to effective symbiotic N_2-fixation in lablab and possibly associative N_2-fixation in finger millet. This merits quantification as do effects of added organic matter in the form of crop residues and manure on soil quality and yields. The RGB and hyperspectral reflectance measurements of crop canopies throughout the season provide the basis for calibrated analyses of high resolution, multi-spectral satellite images, and, thus, an opportunity for the extrapolation of on-station data to the rural–urban interface. Crop water use and leaching data will allow to calculate water use efficiency and leaching and thus also provide an assessment of risks for groundwater contamination.

Acknowledgements This work is carried out as part of the Indo-German Research Unit FOR2432 "Social-Ecological Systems in the Indian Rural-Urban Interface: Functions, Scales, and Dynamics of Transition" jointly funded by the German Research Foundation (DFG) and the Department of Biotechnology (DBT), Government of India ("The Rural-Urban Interface of Bengaluru- a Space of Transitions in Agriculture, Economics, and Society"). Co-funding and infrastructural support has come from the University of Agricultural Sciences Bangalore (UASB), GKVK Campus which is gratefully acknowledged.

References

Astor T, Dayananda S, Nautiyal S et al (2020) Vegetable crop biomass estimation using hyperspectral and RGB 3D UAV data. Agronomy 10(10):1600–1616

Brinkmann K, Hoffmann E, Buerkert A (2020) Spatial and temporal dynamics of urban wetlands along a rural-urban gradient in an Indian megacity over the past 53 years. Remote Sens 12(4):662–695

Diogo RVA, Buerkert A, Schlecht E (2010) Horizontal nutrient fluxes and food safety in urban and peri-urban vegetable and millet cultivation of Niamey, Niger. Nutr Cycl Agroecosyst 87(1):81–102

Friedel JK, Herrmann A, Kleber M (2000) Ion exchange resin-soil mixtures as a tool in net nitrogen mineralisation studies. Soil Biol Biochem 32(11–12):1529–1536

Gericke S, Kurmies B (1952) Die kolorimetrische Phosphorsäurebestimmung mit Ammonium–Vandadat–Molybdat und ihre Anwendung in der Pflanzenanalyse. Z. Düngg. Pflanzenernähr. Bodenk 59:235–247

Jain R, Kishore P, Singh DK (2019) Irrigation in India: status, challenges and options. J Soil Water Conserv 18(4):354–363

Jayakumari R, Nidamanuri RR, Ramiya AM et al (2021) Multi-temporal estimation of vegetable crop biophysical parameters with varied nitrogen fertilization using terrestrial laser scanning. Comp Electron Agric 184:106051

Lompo DJ-P, Sangaré SAK, Compaoré E et al (2012) Gaseous emissions of nitrogen and carbon from urban vegetable gardens of Bobo Dioulasso, Burkina Faso. J Plant Nutr Soil Sci 175(6):846–853

Möckel T, Dayananda S, Nidamanuri RR et al (2018) Multi-temporal biomass estimation of vegetable crops using unmanned aerial vehicles. Remote Sens 10(5):805–823

Murugan R, Parama VRR, Madan B et al (2019) Short-term effect of nitrogen intensification on aggregate size distribution, microbial biomass and enzyme activities in a semi-arid soil under different crop types. Pedosphere 29(4):483–491

Predotova M, Schlecht E, Buerkert A (2010) Nitrogen and carbon losses from dung storage in urban gardens of Niamey, Niger. Nutr Cycl Agroecosyst 87(1):103–114

Swain BB, Teufel N (2017) The impact of urbanisation on crop–livestock farming system: a comparative case study of India and Bangladesh. J Soc Econ Dev 19(1):161–180

VDLUFA (2020) Methodenbuch. https://www.vdlufa.de/Methodenbuch/index.php?lang=de

Werner S, Akoto-Danso E, Manka'busi D et al (2019) Nutrient balances with wastewater irrigation and biochar 3 application in urban agriculture of Northern Ghana. Nutr Cycl Agroecosyst 115(2):249–262

Willich M, Buerkert A (2016) Leaching of carbon and nitrogen from a sandy soil substrate: a comparison between suction plates and ion exchange resins. J Plant Nutr Soil Sci 179(5):609–614

Chapter 8
Effects of Soil Management Practices on Soil Physical Properties and Water Cycle in a Multifactorial Field Experiment

Albara Almawazreh, Daniel Uteau, Andreas Buerkert, Ayappa Sathish, Mudalagiriyappa, D. C. Hanumanthappa, C. T. Subbarayappa, and Stephan Peth

Abstract In the last few decades, Bengaluru's rapid urban expansion and associated population growth brought about profound changes in farming practices, reflected in an increase in irrigated land compared to rainfed land and a growing use of mineral fertilizers. Since it is not clear how intensification of land use due to urbanisation is affecting physical soil properties and field water cycle, we attempt to clarify this at two experimental sites established at the University of Agricultural Sciences Bangalore (UASB). The results show higher bulk densities, lower air capacities, and lower saturated hydraulic conductivity (K_{sat}) values in the rainfed field compared to the irrigated field. Soil moisture was observed to be higher in deeper layers of several plots with low N-level treatments in both experiments. However, it was not significant in other plots. Further statistical investigations and comparisons especially of model

---

A. Almawazreh (✉) · S. Peth
Soil Science, Leibniz University Hannover, 30419 Hannover, Germany
e-mail: albara@ifbk.uni-hannover.de

D. Uteau
Soil Science, University of Kassel, 37213 Witzenhausen, Germany

A. Buerkert
Organic Plant Production and Agroecosystems Research in the Tropics and Subtropics (OPATS), University of Kassel, 37213 Witzenhausen, Germany

A. Sathish
Soil Science and Agricultural Chemistry, University of Agricultural Sciences Bangalore, Bengaluru 560065, India

Mudalagiriyappa
Department of Rainfed Agriculture, University of Agricultural Sciences Bangalore, Bengaluru 560065, India

D. C. Hanumanthappa
University of Agricultural Sciences Bangalore, Bengaluru 560065, India

C. T. Subbarayappa
Soil Science and Agricultural Chemistry, University of Agricultural Sciences Bangalore, Bengaluru 560065, India

E. Hoffmann et al. (eds.), *The Rural-Urban Interface*, The Urban Book Series,
https://doi.org/10.1007/978-3-030-79972-4_8

simulations are required to assess the effect of land use intensifications. Here the experimental approach and some preliminary results are presented.

Keywords Agricultural intensification · Soil management · Physical soil properties · SoilNet · Factorial experiment · Reference data

8.1 Introduction

Over the past two decades, the rapid growth of the south Indian city of Bengaluru and consequently the expansion of urban areas at the expense of rural areas has put major pressure on the region's limited water resources (CGWB 2013). In response, land use and management practices are shifting toward more intense production systems to cope with increasing food demand and changing consumer preferences. This in turn is leading to overexploitation of groundwater resources that are essential for direct human consumption and irrigation.

Currently, there is not much information on how land use intensification and adaption to changes associated with urbanisation in the agricultural ecosystem of the Bengaluru region will influence physical soil properties, which determine water availability and fluxes in the soil–plant-atmosphere continuum (SPAC) on the farm scale. For a better understanding of the water use efficiency, we aimed at modeling the field water fluxes using Hydrus-1D for different treatments (N application, irrigation, crop diversity) established on two experimental fields at UASB—GKVK Campus (Chap. 7). We have parameterized the model and have done a systematic investigation of soil water characteristics, hydraulic functions, and moisture dependent mechanical soil properties.

In this chapter, we describe the methods we have used to estimate hydraulic and physical soil properties based on lab measurements using representative disturbed and undisturbed soil samples and a soil monitoring network for measuring in situ soil water and temperature dynamics.

8.2 Materials and Methods

A multifactorial field experiment was established in July 2016 at GKVK campus, Bengaluru. It consisted of an irrigated and a rainfed field, each with 36 plots (6 m × 12 m) that combined three different crops and three different nitrogen (N) fertilization treatments in four replicates (Chap. 7). During rainy season (August to December), cereals and pulses were cultivated on both fields (maize—*Zea mays* L., finger millet—*Eleusine coracana* Gaertn., and lablab—*Lablab purpureus* (L.) Sweet), and vegetables (tomato—*Solanum lycopersicum* L., eggplant—*Solanum melongena* L., and cabbage—*Brassica oleracea* L.) were cultivated on the irrigated site in the dry season (January to May), while the rainfed field was left fallow. The N fertilizer doses were adjusted to the crop species, applying the recommended dose as medium N-level, the doubled amount as high N-level, and half the amount or zero as low N treatment. Phosphorus and potassium were provided at recommended doses in all treatments. A randomized block design was implemented to assign the treatments at the beginning of the experiment. In subsequent seasons, the crop species were rotated, while the fertilization levels were kept constant; that is, high, medium, and low N-levels were always applied to the same plots, in order to augment the treatment differences over time.

8.2.1 Field Experiments

8.2.1.1 Field Work

Two reference soil profiles were excavated at the middle of both fields (rainfed, irrigated) before the set-up of the multifactorial experiment. Disturbed and undisturbed samples were obtained at 20 cm depth intervals and 15, 40, and 70 cm intervals, respectively. The disturbed samples were used for texture analysis, and the undisturbed samples to determine bulk density (DB) and soil water retention characteristics from which pore size distributions were derived (Fig. 8.1). Soils were classified as Nitisols according to World Reference Base for Soil Resources WRB.

Soil texture was classified by FAO classification as sandy clay loam in the plough horizon and sandy clay in the lower layers (Food and Agriculture Organization 2014).

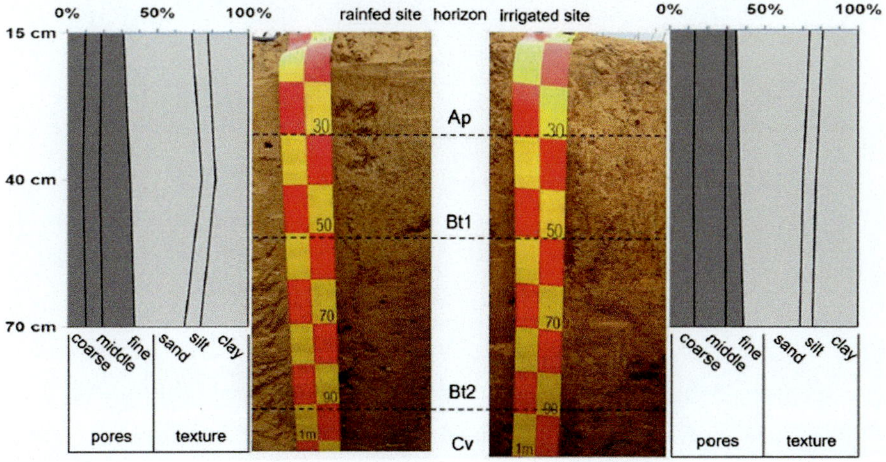

Fig. 8.1 Reference soil profiles and corresponding physical parameters (pore size distribution and texture) measured in 15, 40, and 70 cm depth; values represent means of twelve plots along a central transect of the experimental sites at GKVK Campus, Bengaluru (India); for pore size distribution $n = 36$ for texture $n = 12$

Right after the harvest of the first season crops, between December 2016 and February 2017, all 72 plots in both experiments were sampled in 15, 40, and 70 cm depths for disturbed and undisturbed samples.

From each depth, six soil cores of 4 cm height and 5.5 cm diameter were collected. Six cores were discarded in the rainfed field for having too much rock. In total, 1290 undisturbed samples were used to measure the water retention curve (WRC), DB, and the saturated hydraulic conductivity (K_{sat}) with 12 replicates for each, while 216 disturbed samples were collected at each depth and used to measure texture.

A soil monitoring network was installed at both sites to provide continuous measurements of soil moisture and temperature. Pits were excavated for the installment of the sensors and data loggers using excavators and with the help of field workers. The combined SMT-100 FDR-TDR sensors (SoilNet, FZ Jülich, Germany) were placed at 15, 40, and 70 cm depths in each plot in both experiments (108 sensors each).

Sensor measurements are read in 10 min intervals and collected by 24 data loggers (12 for each experiment) and then transmitted from each experiment to a receiver placed on the roof of a building located between the two sites (Fig. 8.2). The SoilNet is in operation since February 2017 and the monitoring of soil moisture and temperature will last for at least six years.

An ultrasonic water flow sensor was installed at the irrigated experiment to monitor the water volume used for irrigating the crops and is in operation since February 2018.

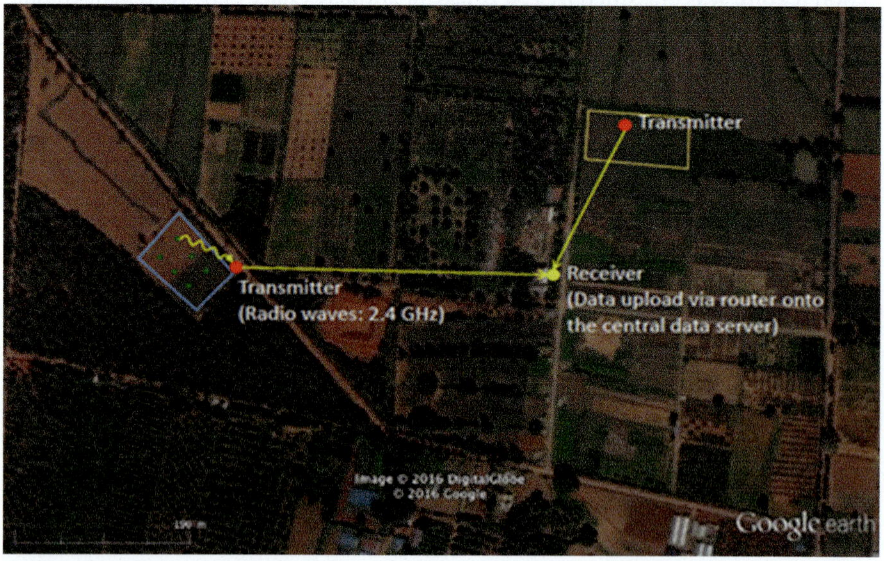

Fig. 8.2 Map of the two experimental sites: rainfed on the upper right and irrigated on the left. Transmitters send the data to the receiver on the roof of a building in the middle at GKVK Campus, Bengaluru (India)

Fig. 8.3 Customized falling head permeameter to measure saturated hydraulic conductivity (left) and suction plate assembly to determine water retention functions (middle and right) that were built at GKVK Campus, Bengaluru (India)

8.2.1.2 Laboratory Work

A customized pressure plate assembly is used to measure the water retention curve, and a falling head permeameter was constructed for measuring saturated hydraulic conductivity K_{sat} (Fig. 8.3).

Samples of the water retention curves were measured at matric potentials of −60, −150, −300, −500 and −15,000 hPa. Those results are used to determine the soil hydraulic properties in the Hydrus-1D model. Results were also used to calculate the permanent wilting point (PWP), total pore volume (PV), air capacity (AC), and plant available water (PAW).

8.3 Results and Discussion

Due to the different management history of the fields and differences in soil texture (36% clay at the irrigated and 27% clay at the rainfed experiment), soil physical properties varied between the two sites. In the rainfed experiment, higher bulk densities (15 cm: 1.69 g/cm^3, 70 cm: 1.55 g/cm^3) were observed compared to the irrigated site (15 cm: 1.65 g/cm^3, 70 cm: 1.52 g/cm^3). Thus, total pore volume in the rainfed site was about 5 vol% smaller than in the irrigated field. Higher air capacities corresponded with higher K_{sat} values in the irrigated field.

Comparing the physical properties of both sites with respect to water supply, fertilizer treatment and crops, the following tendencies were observed: K_{sat} values ranged for both experiments between low and medium (according to AG Boden 2005) and differed significantly at 15 cm and 40 cm soil depth with higher values in the irrigated experiment (Fig. 8.4). With respect to crop effects and fertilizer level, no significant differences were observed between both experimental sites (data not shown). Air capacities values were significantly lower in the irrigated experiment (<5 vol%) which is critical for soil aeration. The same is true for plant available water which was significantly lower in the rainfed experiment (<15 vol%). As sampling

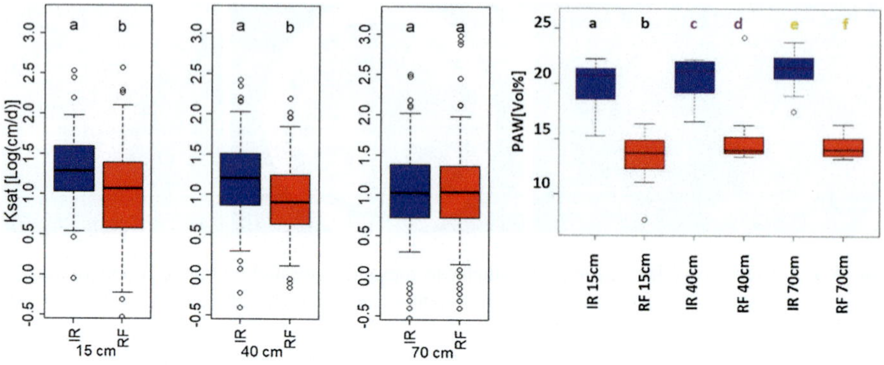

Fig. 8.4 Boxplots of saturated hydraulic conductivity (left) and plant available water (right) in depths of 15, 40, and 70 cm for irrigated (IR) and rainfed (RF) experiment ($n = 108$) at GKVK Campus, Bengaluru (India). Significant difference is indicated by different letters (Fuchs 2017, unpublished)

was done after just one cropping season, results reflect a very early stage of the experiments.

The monitoring of soil moisture by the SMT-100 sensors continued from the second cropping season onwards. In the rainfed field, no response to the different nitrogen fertilization treatments was found so far regarding the soil moisture content. While in some plots, soil moisture content in the deeper layers seemed to be lower for the high N-level in comparison to low N (Fig. 8.5); on other plots, the opposite was observed. Here soil moisture in the root zone was higher in the high N treatment (Fig. 8.6).

This might suggest that in those fields, macropores are connecting deeper layers with the surface, and thereby, water is being transported faster leading to higher moisture content observations. The same can be stated about the irrigated field.

As expected, under constant irrigation soil moisture in the irrigated experimental field remained at a higher level than under rainfed conditions (Fig. 8.7). These preliminary observations will be analyzed in more detail once all currently available monitoring data sets are processed and cross-correlated with crop yield data (Chap. 7).

The comprehensive, high resolution time series of soil moisture and temperature measurements are currently used to calibrate and validate the hydraulic model which will subsequently be used to solve the field water balance and to calculate water use efficiencies of high and low input farming systems.

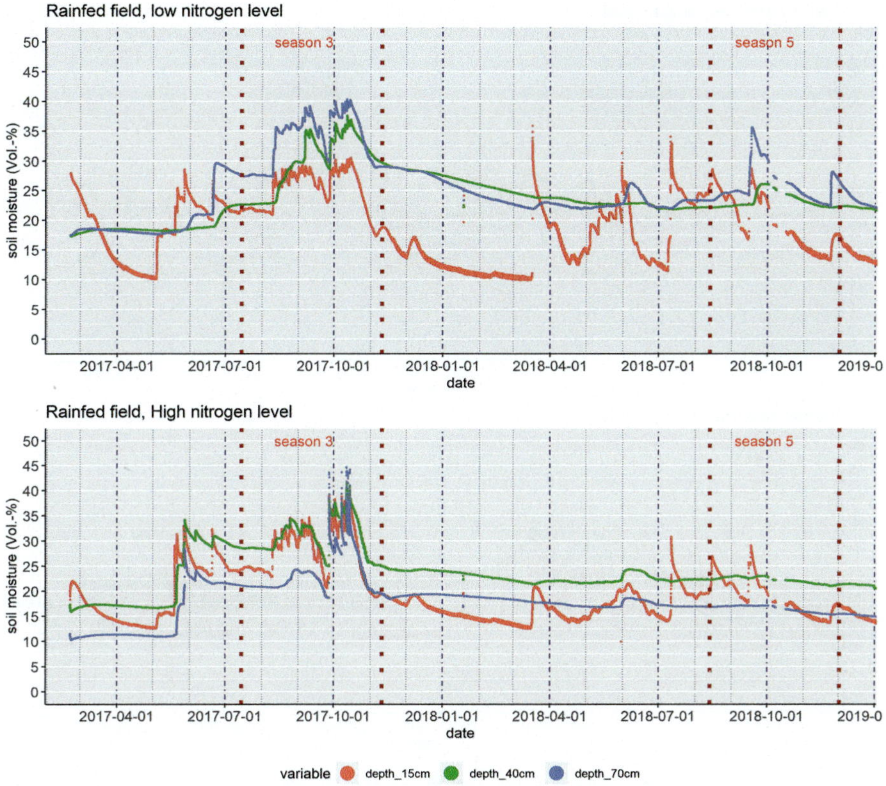

Fig. 8.5 Soil moisture in depths of 15, 40, and 70 cm, monitored by the SoilNet sensor network, shown for finger millet (season 3) and lablab (season 5), in plots 11 and 10, respectively, of a factorial field experiment at GKVK Campus, Bengaluru (India). Brown dotted lines indicate sowing and harvesting

8.4 Conclusion

After one season, the different crops and fertilizer levels did not seem to have a tangible impact on K_{sat}. Yet, there were differences in the soil's physical properties between the fields that will be considered in the Hydrus-1D model. The results from the soil moisture sensors confirm the higher K_{sat} in the irrigated field compared with the rainfed field where higher soil moisture was observed in the whole soil profile.

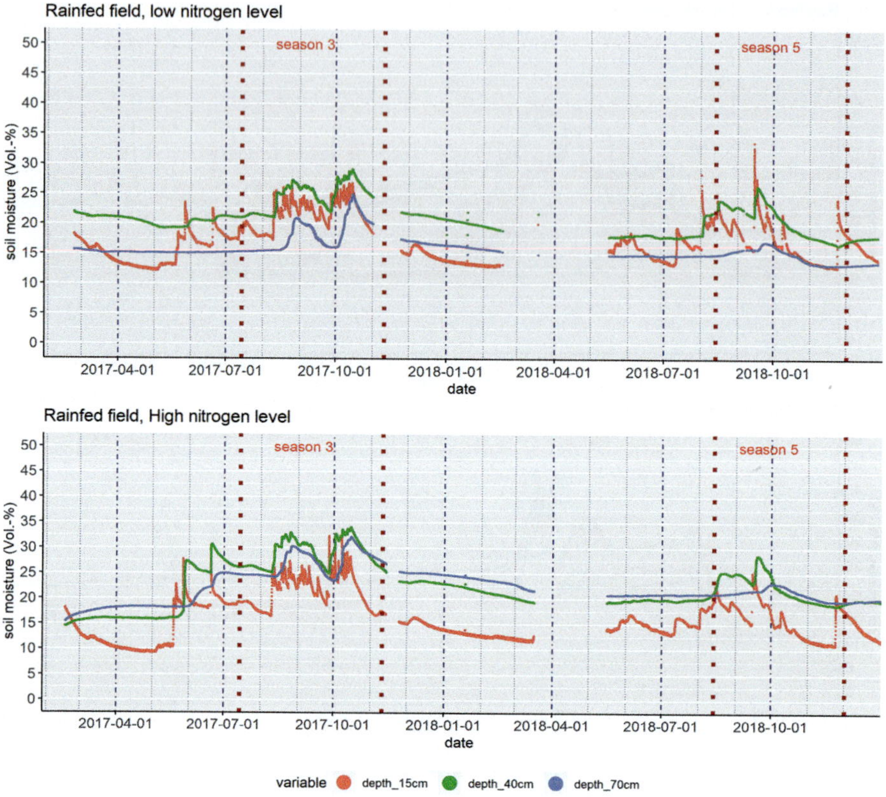

Fig. 8.6 Soil moisture in depths of 15, 40, and 70 cm, monitored by the SoilNet sensor network, shown for finger millet (season 3) and lablab (season 5), in plots 19 and 21, respectively, of a factorial field experiment at GKVK Campus, Bengaluru (India). Brown dotted lines indicate sowing and harvesting

While first soil moisture data indicate effects of different N-levels for some plots in both experiments, this trend needs further confirmation. The results from our calibrated hydraulic model will allow comparing individual components of the water cycle, particularly groundwater recharge, plant water uptake, and water use efficiencies for different crops at three intensities of N.

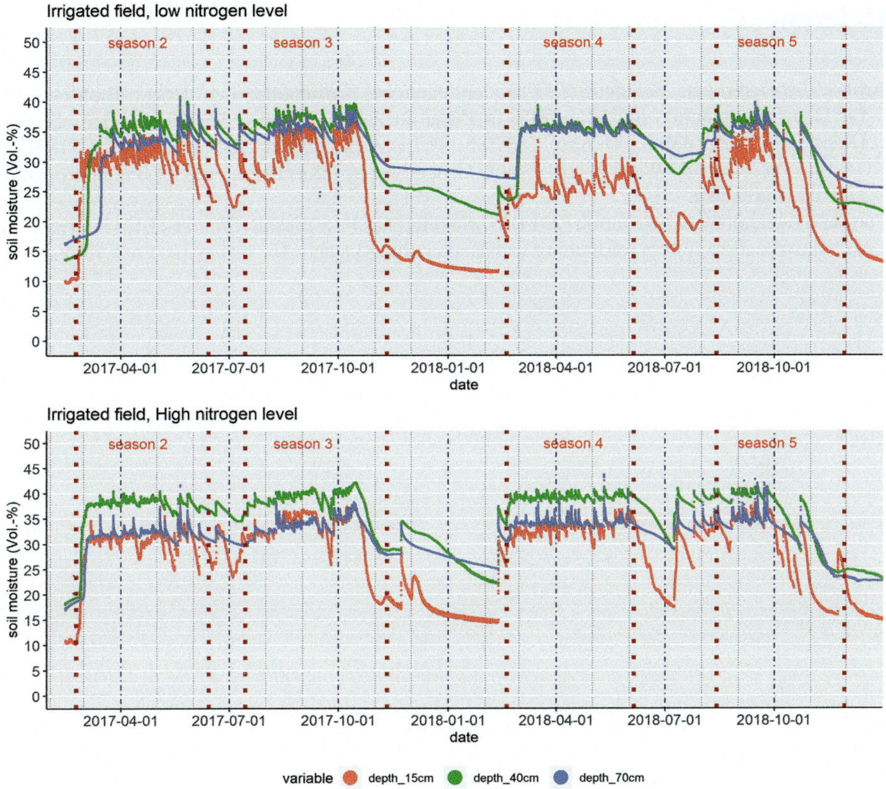

Fig. 8.7 Soil moisture in depths of 15, 40, and 70 cm, monitored by the SoilNet sensor network, shown for tomato, maize, cabbage, and finger millet (seasons 2–5), in plots 19 and 21, respectively, of a factorial field experiment at GKVK Campus, Bengaluru (India). Brown dotted lines indicate sowing and harvesting

Acknowledgements We would like to thank Dipl.-Geoökol. Sybille Lehmann and BSc. Michael Tobias Fuchs for their contributions to installing the sensor network, sampling and laboratory analysis. This study is funded by the DFG (German Research Foundation) in the frame of the research unit FOR2432 (Social-Ecological Systems in the Indian Rural-Urban Interface: Functions, Scales, and Dynamics of Transition) under project number 279374797.

References

Ad-hoc-Arbeitsgruppe Boden (2005) Bodenkundliche Kartieranleitung, Hrsg.: Bundesanstalt für Geowissenschaften und Rohstoffe in Zusammenarbeit mit den Staatlichen Geologischen Diensten, 5. Aufl., 438 S.; 41 Abb., 103 Tab., 31 Listen, Hannover 2005. ISBN 978-3-510-9592

CGWB Central Ground Water Board (2013) Ground water information booklet Bangalore Urban District, Karnataka.

Food and Agriculture Organization (2014) World reference base for soil resources 2014 international soil classification system for naming soils and creating legends for soil maps. FAO, Rome

Chapter 9
Biomass Estimation of Vegetables—Can Remote Sensing Be a Tool for It?

Thomas Astor and Michael Wachendorf

Abstract Precision agriculture for vegetable crops has received little interest so far, although remote sensing (RS) has been used as an effective tool to monitor other agricultural production systems. The aim of this study therefore was to assess the applicability of RS imagery for capturing crop height information of three vegetable crops (eggplant, tomato, cabbage) during a complete growth cycle to infer biomass. Additionally, the effect of crop development stage on the relationship between estimated crop height and field measured crop height was examined. The study was conducted in an experimental layout at the University of Agricultural Sciences Bangalore in Bengaluru, India. Crop height and the biomass were measured at five dates during the crop growth cycle between February and May 2017. The prediction accuracy varied strongly depending on the applied statistical approach. For all crops, a good predictive performance with a cross-validated prediction error <20% was achieved. The growth stage of the crops had no significant effect on the prediction accuracy, although increasing trends of an underestimation of fresh matter biomass at later sampling dates were found for eggplant and tomato. The study proves that the estimation of vegetable biomass using RS data is successful throughout the growing season.

Keywords Vegetable crops · Crop phenology · Biomass estimation · Yield · Factorial experiment · Remote sensing · Hyperspectral imaging · Machine learning

---

T. Astor (✉) · M. Wachendorf
Grassland Science and Renewable Plant Resources, University of Kassel, 37213 Witzenhausen, Germany
e-mail: thastor@uni-kassel.de

M. Wachendorf
e-mail: mwach@uni-kassel.de

E. Hoffmann et al. (eds.), *The Rural-Urban Interface*, The Urban Book Series,
https://doi.org/10.1007/978-3-030-79972-4_9

9.1 Introduction

The development of megacities by rapid urbanisation, in India and elsewhere, poses increasing pressure on agricultural production systems (Rockström et al. 2009). To meet the increasing pressure of population growth in combination with widespread malnutrition especially in developing countries, sustainable agriculture is necessary in order to increase food production (Tilman et al. 2011). For decades (Tucker, 1979; Pettorelli et al. 2005), remote sensing (RS) has been an effective tool to monitor agricultural production systems (Atzberger 2013; Duncan et al. 2015). Particularly in countries with several cropping seasons per year, such as India, agricultural systems require multi-temporal RS data for a continuous evaluation of crop phenology (Mondal et al. 2014). The evaluation of the crop status can help farmers to identify deficiencies due to lack of water or infestation by pests early before the damage to the crop gets serious. Particularly, vegetable production is characterized by frequent nitrogen (N) fertilization (Thompson et al. 2018) and requires a rapid assessment of crop status (Padilla et al. 2018).

For more than two decades precision agriculture (PA) has been applied for field crops, but in horticulture PA is so far only poorly applied (Zude-Sasse et al. 2016). The smaller field size and the reduced planting density as well as the complex plant architecture makes PA in horticulture a complex task (Zude-Sasse et al. 2016). Recently, unmanned aerial vehicles (UAV) have gained interest as platforms for collecting RS information for horticulture (Duarte-Carvajalino et al. 2018; Moeckel et al. 2018). UAVs allow collection of data from large areas in a highly repetitive manner, which makes them more independent from weather conditions than satellite or airborne platforms (Zhang and Kovacs 2012).

Remote sensing data are available from a range of sensors. Each sensor measurement system has its own strengths and limitations. Sensor fusion may allow a more holistic interpretation of the relationship between RS information and crop parameters. For example, the fusion of spectral reflectance information with data from an ultra-sound system improved the prediction performance for barley yield (Rischbeck et al. 2016). Other studies have shown higher prediction precision for vegetation parameters (e.g., leaf area index and biomass) of crops, such as maize (*Zea mays* L.; Gao et al. 2013), winter wheat (*Triticum aestivum* L.; Yue et al. 2018), and soybean (*Glycine max*. (L.) Merr.; Maimaitijiang et al. 2017), when spectral and structural information (based on RADAR, LiDAR, or photogrammetric methods) were combined. For horticulture, however, there is a lack of research on the benefits of sensor integration.

The primary objective of this study was to test the potential of two types of RS data—RGB (red–green–blue) data acquired by a UAV and terrestrial hyperspectral data—to predict fresh matter biomass (FMB) of eggplant (*Solanum melongena* L.), tomato (*Solanum lycopersicum* L.), and cabbage (*Brassica oleracea* L.). For establishing the empirical relationship, four different machine learning (ML) approaches were tested (partial least squares, support vector machines, random forest, and gradient boosting trees regression). The contribution and potential of data from each sensor individually (crop height values from the RGB dataset and spectral information from the hyperspectral dataset) and from both sensors fused together were evaluated.

9.2 Materials and Methods

The data for this study were collected on the experimental station of the University of Agricultural Sciences Bangalore (UASB) in Bengaluru, India (12° 58′ 20.79″ N, 77° 34′ 50.31″ E, 920 asl). The dominant soil type is a Kandic Paleustalf or Dystric Nitisol. The site is characterized by a tropical climate with a distinct monsoon season from June to October, a mean annual temperature of 29.2 °C, and an average annual precipitation of 923 mm (Prasad et al. 2016).

The field experiment was established in 2016 (Chap. 7) and vegetables were cultivated in the dry season (January to May) using a drip irrigation system. Three crop species, tomato (*Solanum lycopersicum* L. cv. NS-501), eggplant (*Solanum melongena* L. cv. Ankur), and cabbage (*Brassica oleracea* L. cv. Unnati) were grown with three levels of nitrogen (N) fertilizer in four repetitions. The medium N rate was the recommended N dose (50 kg N ha^{-1} for eggplant, 46 kg N ha^{-1} for tomato, and 60 kg N ha^{-1} for cabbage), whereas the high (+50%) and low (−50%) dose represented additional intensities typical in the study area. Recommended doses of phosphorus and potassium were provided in a blanket application (40 kg P ha^{-1} and 20, 24, and 50 kg K ha^{-1} for eggplant, tomato, and cabbage). As fertilizer was directly applied to individual plants, no nutrient deficiency was noticeable in any of the treatments. The plots (6 m × 12 m) were divided into a sub-subplot for destructive biomass sampling (S-plot) and a sub-subplot for non-destructive spectral measurements (H-plot) of 6 m × 6 m each.

The total biomass (vegetative parts as well as fruits at later growth stages) was measured at five times between March and May 2017. At each sampling date, two randomly selected plants were harvested 2 cm above the ground in each S-plot, their biomass was measured in the field, and extrapolated to the plot area. The total number of samples collected for each crop type was 60 (5 sampling dates × 3 fertilizer treatments × 4 replicates).

At each sampling date, one RGB UAV flight for each plot was conducted using a DJI FC300X camera (focal length of 3.61 mm) mounted on DJI 3 Professional drone (DJI, Shenzhen, Guangdong, China). The same grid-wise flight plan was conducted for each plot using autopilot and the software Pix4D Capture (Pix4D SA, Lausanne, Switzerland). About 200 images were taken during each flight with an 80% forward and side overlap. The flight altitude was set to 20 m (ranging from 17 to 24 m, due to variations in the built-in GPS of the drone) and the mean ground resolution was ranging between 6 and 8 mm per pixel. As precise point cloud processing requires high spatial position accuracy, four permanent ground control points (GCPs) were measured in each main plot using a differential GPS (Trimble Inc., Sunnyvale, CA,

USA) with a spatial accuracy of ~5 cm. These GCPs were subsequently used for geo-referencing, geo-correction, and co-registration of the images. Point cloud data were derived from the UAV RGB images using the structure-from-motion algorithm from the software Photoscan Pro (AgiSoft LLC, St. Petersburg, Russia). With the help of the RGB information, all points were classified into ground and vegetation points. With the resulting ground points, a digital elevation model (DEM) was created with a spatial resolution of 1 m. The DEM was used to normalize the height information for all points classified as vegetation. A total of 13 point cloud metrics were derived from the vegetation point cloud, which were used as explanatory variables from the height dataset for the biomass estimation: minimum crop height, maximum crop height, mean crop height, standard deviation of crop height, median crop height, skewness of crop height, kurtosis of crop height, coefficient of variation crop height, 70th percentile of crop height, 80th percentile of crop height, 90th percentile of crop height, 95th percentile of crop height, 99th percentile of crop height, and crop canopy relief height. The extraction of the point cloud was done using the lidR package (Roussel and Auty 2017) and the program R (R Core Team 2016). Prior to the biomass sampling, three hyperspectral images were collected in each H-plot using the full-frame hyperspectral camera UHD 185-Firefly (UHD; Cubert GmbH, www.cubert-gmbh.de). Each image was collected above a randomly chosen plant. The distance between the plant and the camera was kept at 1.5 m throughout the growing season to keep the area covered by the image during plant growth constant. The camera measured the reflectance in the spectral range from 470 to 950 nm. The band width was 4 nm, resulting in 120 bands which were used as explanatory variables from the spectral dataset. The camera was mounted on a tripod to reduce the effects of shading and movement by the human operator as well as to keep the measurement height constant. The reflectance for each image was normalised by dividing the reflectance information for each pixel by the maximum measured reflectance value for this pixel. Each collected image contained non-vegetation pixels, such as soil, shadows, and pipes from the drip irrigation system. To reduce the effect of such disturbances, a stepwise procedure was applied. In the first step, the normalized vegetation index (Rouse et al. 1974) was calculated for each pixel of the image. Subsequently, a k-mean classification was applied to distinguish vegetated from non-vegetated surfaces. Finally, the average reflection for each image was calculated based on all vegetated pixels in an image. All images collected in the same plot and at the same sampling date were averaged. This process results in one spectral reflectance curve per plot and sampling date.

Four machine learning (ML) techniques, i.e., partial least squares (PLS), support vector machines (SVM), random forest (RFR), and gradient boosting tree (GBT) regression were applied to calibrate the biomass prediction models. ML methods are empirical statistical approaches which can be used to establish statistical relationships between a multitude of explanatory variables (here: spectral bands and height parameters) and response variables (here: crop biomass). To assure a proper calibration and validation procedure, the datasets for each crop ($n = 60$) were split up into a calibration dataset ($n = 45$) and a validation dataset ($n = 15$). As all ML methods can be optimized by tuning a certain parameter (e.g., number of latent variables for PLS), an optimisation procedure at the model calibration of all four ML approaches was done using tenfold group-leave out cross-validation. For this, 40 samples were used for the model calibration, while the remaining five samples were used for model validation. This was repeated ten times, and the optimal parameters for the calibration model were determined subsequently based on the lowest root-mean-square-error (RMSE).

9.3 Results

In total, 36 ML models were calibrated and validated. The prediction accuracy for the three crops varied strongly depending on the calibration model applied and dataset used (eggplant: $R^2_{val} = 0.77 - 0.97$, nRMSE = 0.07 − 0.18; tomato: $R^2_{val} = 0.59 - 0.97$, nRMSE = 0.06 − 0.23; cabbage $R^2_{val} = 0.76 - 0.93$, nRMSE = 0.09 − 0.17). Taylor diagrams (Taylor 2001) were used to evaluate the precision of the biomass models as they provide a quick and easy visual comparison of the predicted versus observed values for several models. For this purpose, the Pearson correlation coefficient (r), the root-mean-square difference (RMSD), and the standard deviation of the predicted and observed values of the validation dataset were calculated for each of the 12 models for each crop (Fig. 9.1). Only the Taylor diagrams for tomato showed a larger variation of the RMSD (~0.1 − 0.5) and r (~0.8 − 0.99) values for all models (Fig. 9.1), indicating dependencies of the selected ML approach and the used dataset for predicting tomato biomass. For eggplant, only PLS regression using the height parameters showed a lower predictive performance (based on RMSD and r) than the other models (Fig. 9.1a). The models predicting cabbage biomass all showed a similar predictive performance (Fig. 9.1c). For all three crops, RFR showed

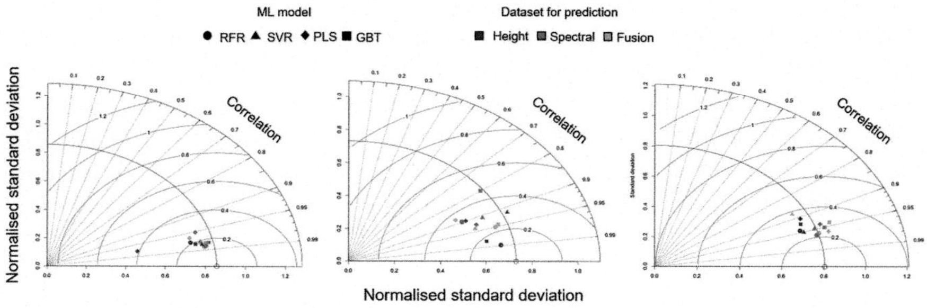

Fig. 9.1 Taylor diagrams displaying a statistical comparison of the 12 biomass estimation models and the observed data of the validation dataset for **a** eggplant (left), **b** tomato (center), and **c** cabbage (right). The best model is located nearest to the observed point (•) on the x-axis, while also having a relatively high-correlation coefficient (dotted lines) and low values for the root-mean-square difference (green lines)

the best results, with an R^2_{val} of 0.97 for eggplant and tomato and 0.93 for cabbage. The relative prediction error (nRMSE) was largest for cabbage (9%), followed by eggplant (7%) and tomato (6%).

9.4 Discussion

Overall, the results show that a high prediction accuracy of FMB can be achieved for vegetable crops with the chosen methods (nRMSE: 6% for tomato, 7% for eggplant and 9% for cabbage). The presented direct approach (i.e., predicting biomass directly from remotely sensed variables) resulted in much more accurate biomass estimates than indirect approaches (Madec et al. 2017; Moeckel et al. 2018).

Among the four tested ML approaches, RFR performed best for all three crops. As RFR is less sensitive to overfitting than many other methods (Belgiu and Drăguţ 2016) RFR frequently outperforms other ML methods in predicting biomass from RS data in agriculture (Moeckel et al. 2018; Nasi et al. 2018). Nevertheless, a thorough evaluation of several ML methods should always be conducted, as there is no ML method, which consistently outperforms the others under all circumstances such as ecosystem types and vegetation parameters (Flach 2012).

To prove whether the fusion of spectral and height information can improve biomass predictions in comparison to single sensor use, biomass estimation models, based on spectral data and point cloud-derived height information alone and in fusion, were developed. The sensor data fusion did not result in the optimal performance for any of the crops. This could partly be explained by data redundancy when using both datasets as crop height and reflectance in the near-infrared spectral region (760–950 nm) were strongly correlated with vegetation biomass (Pohl and van Genderen 1998). While for eggplant and cabbage, the spectral dataset delivered the best prediction accuracies, the model based on the height parameters was the best for tomato. The agricultural practice of tying up the tomato plants from the third sampling date onward directly affected the height measurements and thus the point cloud derivation and might be an explanation for the selection of the height parameters. These results confirm the expectation that an evaluation of two or more RS sets of variables reflecting different vegetation characteristics (i.e., spectral reflectance and structure) is as important (Prošek and Šímová 2019) as evaluating different ML methods (Nasi et al. 2018).

9.5 Conclusions

In future, this approach will be tested using a completely airborne data collection to validate the findings on larger scales. The approach can then be used to evaluate spatial variability of crop biomass under field conditions. This step would allow further development of precision agriculture in horticultural systems as it would allow evaluating phenotypic characteristics of vegetable crops on large scales in a reliable way. As a diagnostic tool, the methodology can serve to determine crop management intensification on farmer fields (up to village and regional scales), while avoiding tedious point sampling on the ground, and thus, support the real-time observation of agricultural transitions along the rural–urban gradient.

Acknowledgements The authors gratefully acknowledge the help of Dr. Supriya Dayananda for collecting the data in the field, Prof. Dr. Andreas Buerkert for providing the RGB drone used in this study as well as Nagaraju Hanumaiah for running the experimental site within the University of Agricultural Sciences Bangalore in Bengaluru. The authors acknowledge the financial support provided by the German Research Foundation, DFG, through grant numbers WA 2135/4-1 and BU 1308/13-1 and by the Indian Department of Biotechnology, DBT, through grant number DBT/IN/German/DFG/14/BVCR/2016 as parts of the Indo-German consortium of DFG Research Unit FOR2432/1 and DBT (The Rural-Urban Interface of Bangalore: A Space of Transitions in Agriculture, Economics, and Society).

References

Atzberger C (2013) Advances in remote sensing of agriculture: context description, existing operational monitoring systems and major information needs. Remote Sens 5(2):949–981

Belgiu M, Drăgut L (2016) Random forest in remote sensing: a review of applications and future directions. ISPRS J Photogramm Remote Sens 114:24–31

Duarte-Carvajalino JM, Alzate DF, Ramirez AA et al (2018) Evaluating late blight severity in potato crops using unmanned aerial vehicles and machine learning algorithms. Remote Sens 10(1513):1–17

Duncan JMA, Dash J, Atkinson PM (2015) The potential of satellite-observed crop phenology to enhance yield gap assessments in smallholder landscapes. Front Environ Sci 3(56):1–16

Flach P (2012) Machine learning: the art and science of algorithms that make sense of data. Cambridge Univ. Press, Cambridge

Madec S, Baret F, Solan B et al (2017) High-throughput phenotyping of plant height: comparing unmanned aerial vehicles and ground LiDAR estimates. Front Plant Sci 8(2002):1–14

Maimaitijiang M, Ghulam A, Sidike P et al (2017) Unmanned Aerial System (UAS)-based phenotyping of soybean using multi-sensor data fusion and extreme learning machine. ISPRS J Photogramm Remote Sens 134:43–58

Moeckel T, Dayananda S, Nidamanuri RR et al (2018) Estimation of vegetable crop parameter by multi-temporal UAV-borne images. Remote Sens 10(5)

Mondal S, Jeganathan C, Sinha NK et al (2014) Extracting seasonal cropping patterns using multi-temporal vegetation indices from IRS LISS-III data in Muzaffarpur District of Bihar, India. Egypt J Remote Sens Space Sci 17(2):123–134

Nasi R, Viljanen N, Kaivosoja J et al (2018) Estimating biomass and nitrogen amount of barley and grass using UAV and aircraft based spectral and photogrammetric 3D features. Remote Sens 10(7):1–32

Padilla FM, Gallardo M, Peña-Fleitas MT (2018) Proximal optical sensors for nitrogen management of vegetable crops: a review. Sensors 18(7):1–23

Pettorelli N, Vik JO, Mysterud A et al (2005) Using the satellite-derived NDVI to assess ecological responses to environmental change. Trends Ecol Evol 20(9):503–510

Pohl C, van Genderen JL (1998) Review article multisensor image fusion in remote sensing: concepts, methods and applications. Int J Remote Sens 19(5):823–854

Prasad JVNS, Rao CS, Srinivas K (2016) Effect of ten years of reduced tillage and recycling of organic matter on crop yields, soil organic carbon and its fractions in Alfisols of semi arid tropics of southern India. Soil Tillage Res 156:131–139

Prošek J, Šímová P (2019) UAV for mapping shrubland vegetation: does fusion of spectral and vertical information derived from a single sensor increase the classification accuracy? Int J Appl Earth Obs Geoinf 75:151–162

Rischbeck P, Elsayed S, Mistele B et al (2016) Data fusion of spectral, thermal and canopy height parameters for improved yield prediction of drought stressed spring barley. Eur J Agron 78:44–59

Rockström J, Steffen W, Noone K et al (2009) A safe operating space for humanity. Nature 461(7263):472–475

Rouse JW, Haas RH, Schell JA et al (1974) Monitoring vegetation systems in the Great Plains with ERTS. NASA Spec Publ 351:309

Roussel JR, Auty D (2017) lidR: Airborne LiDAR data manipulation and visualization for forestry applications. R package version 1(1). https://CRAN.R-project.org/package=lidR

Thompson RB, Voogt W, Incrocci L et al (2018) Strategies for optimal fertiliser management of vegetable crops in Europe. In: V international symposium on ecologically sound fertilization strategies for field vegetable production, vol 1192, pp 129–140

Tilman D, Balzer C, Hill J (2011) Global food demand and the sustainable intensification of agriculture. Proc Natl Acad Sci USA 108:20260–20264

Tucker CJ (1979) Red and photographic infrared linear combinations for monitoring vegetation. Remote Sens Environ 8(2):127–150

Yue J, Feng H, Jin X et al (2018) A comparison of crop parameters estimation using images from UAV-mounted snapshot hyperspectral sensor and high-definition digital camera. Remote Sens 10(7):1–24

Zhang C, Kovacs JM (2012) The application of small unmanned aerial systems for precision agriculture: a review. Precision Agric 13(6):693–712

Zude-Sasse M, Fountas S, Gemtos TA et al (2016) Applications of precision agriculture in horticultural crops. Eur J Hortic Sci 81(2):78–90

Chapter 10
Dairy Farms in an Urbanising Environment: Tradeoffs Between Productivity and Animal Welfare

Ana Pinto, Tong Yin, Marion Reichenbach, Pradeep Kumar Malik, Eva Schlecht, and Sven König

Abstract The aim of the present study was to analyse daily milk yield and cow leg health considering farms located in urban, peri-urban and rural areas. The study allows deeper insight into processes of urbanisation and their impact on dairy cattle farming in Bengaluru, a rising megacity in India with currently more than 11 million inhabitants. Urban dairy cattle farming operates under several constraints regarding cow welfare aspects, i.e. limited grazing access, sub-optimal manure management and partly poor housing conditions with limited space. The poor housing conditions, especially in combination with larger herd sizes and concrete floor, might explain the higher disease incidences for hock disorders in urban areas. Despite the environmental constraints and leg problems, cow productivity in terms of milk yield was higher in urban than in rural areas. The priorisation of productivity is likely due to economic drivers, foremost the increasing demand for dairy products in urban districts, combined with limited competition and high milk prices.

Keywords Livestock husbandry · Dairy cattle · Farm assessment · Genetic diversity · Functional trait patterns · Productivity

---

A. Pinto (✉) · T. Yin · S. König
Animal Breeding and Genetics, Justus-Liebig-University Gießen, 35390 Gießen, Germany
e-mail: ana.pinto-garcia@agrar.uni-giessen.de

T. Yin
e-mail: tong.yin@agrar.uni-giessen.de

S. König
e-mail: sven.koenig@agrar.uni-giessen.de

M. Reichenbach · E. Schlecht
Animal Husbandry in the Tropics and Subtropics, University of Kassel, Witzenhausen and Georg-August-University Göttingen, 37213 Göttingen, Germany
e-mail: marion.reichenbach@uni-kassel.de

E. Schlecht
e-mail: eschlech@uni-kassel.de

P. K. Malik
ICAR-National Institute of Animal Nutrition and Physiology, Bengaluru 560030, India
e-mail: malikndri@gmail.com

E. Hoffmann et al. (eds.), *The Rural-Urban Interface*, The Urban Book Series,
https://doi.org/10.1007/978-3-030-79972-4_10

10.1 Introduction

Urbanisation processes are associated with rapid economic growth (Chao and Dongmei 2015), contributing to agricultural intensification. Bengaluru is the third most populous city in India, and the number of its inhabitants has doubled during the past 15 years. According to Oxford Economics (2018), the fastest annual economic growth worldwide for the years 2019–2035 is expected for Bengaluru (8.5% year-by-year growth of gross domestic product), contributing to the increasing economic strength of India.

Moreover, most of the vegetarians worldwide are located in India. In 2014, percentages of vegetarianism were 28.4 in male and 29.3 in female inhabitants (Office of the Registrar General & Census Commissioner, India 2014), implying that dairy products are the most important protein source. In consequence, the population growth accompanied by increasing income and an increasing demand for milk products forces the dairy industry and dairy cattle farmers to enhance production efficiency. However, increasing production efficiency in urban areas is associated with further difficulties, such as limited grazing areas, fodder scarcity, problems of manure disposal, and sub-optimal cattle husbandry conditions. Furthermore, rapid urbanisation might influence dairy cattle welfare and health traits (de Vries et al. 2011). Hence, the objective of the present study was a detailed evaluation of alterations of cow traits along social-ecological gradients in the context of urbanisation processes in Bengaluru.

10.2 Materials and Methods

A total of 582 dairy cattle were selected from 31 different villages located along the northern and southern transects of Bengaluru (Chap. 1), as well as the central city. Following the SSI classification described by Hoffmann et al. (2017), every village received a value ranging from 0 to 1, and the villages were clustered in the three areas, urban (SSI < 0.3), rural (SSI > 0.5), and peri-urban or transitional (SSI: 0.3–0.5). For cow trait recording, 121 dairy cattle farms were visited three times in four-month intervals from June 2017 to April 2018. Herd size ranged from 2 to 12 lactating cows, and the average lactation number was 2.51.

Trait recording included daily milk yield (MY, in litre) as production trait and the hock assessment score (HAS) as a functional trait. For HAS, possible scores were 1, 2 or 3. According to Lombard et al. (2010), score 1 was assigned for hocks with hair and without swelling; score 2 for hocks without hair and without swelling and score 3 for hocks with hair loss and swelling. Only one trained person was responsible for all trait recordings. Furthermore, breed, lactation number and days in milk were recorded as explanatory variables.

Table 10.1 Distribution of breeds and traits of dairy cattle in the rural–urban interface of Bengaluru

	Urban	Transitional	Rural
Breeds			
HF (%)	33	47	54
Jersey (%)	16	8	10
Native (%)	4	7	3
Crossbreds (%)	47	38	33
Traits			
MY (l/day)	9.49	8.92	8.01
HAS	1.65	1.50	1.49

Linear mixed models implemented in R (Bates et al. 2015) were applied to analyse the trait pattern for MY and HAS in dependence of SSI. Farm and repeated measurements for cow effects were modelled as random effects, and breed, SSI, days in milk, season, and lactation number were considered as classification effects.

10.3 Results

Significant interaction (p-value < 0.05) was found for both, MY and HAS. In the case of MY, least squares means decreased along the SSI. Cows from the urban area had a higher productivity (MY 9.49 l/day) compared with cows from the rural area (MY 8.01 l/day; Table 10.1). Hence, cows from urban farms were more productive than cows from rural farms. Least squares means for HAS decreased gradually from urban (1.65 score) to rural (1.49 score) districts, meaning that cattle in rural areas had better hock condition and less injuries than cattle in urban areas.

Independent from the location (urban, peri-urban or rural), generally the most frequent breed in the northern as well as in the southern transect was Holstein Frisian (HF), followed by Jersey (Table 10.1). In urban areas, an increasing percentage of crossbred cattle was observed (urban: 47% crossbred cattle; rural: 33% crossbred cattle). The percentage of HF was higher in rural areas (urban: 33% HF cattle; rural: 54% HF cattle).

10.4 Discussion

Highest productivity in terms of MY was determined for cows from urban areas. Urbanisation has increased the market demand for milk and milk products, although urban and peri-urban dairy farms are under great pressure with regard to fodder scarcity, labour shortage and inappropriate disposal of animal excreta (Chap. 5).

Fig. 10.1 Pictures from dairy cattle sheds in Bengaluru, India. Left, dairy cattle are kept on a ground floor of a house in an urban area. Right, cattle are kept in an outside shed in a rural area. The shed is built with stone columns and palm leaves as a roof

However, farmers in urban areas can sell their milk directly to customers and decide their own price (31–40 INR/liter[1]). In rural areas, milk collection centres pay lower and fixed prices ranging from 23 to 27 INR/liter[2] depending on the milk fat content. Hence, the higher milk production in urban farms is stimulated by the higher milk prices.

However, higher MY in urban herds was associated with higher HAS. According to Rutherford et al. (2008), incidences of hock lesions (i.e. impaired cow health) might be due to increasing productivity, implying metabolic stress and causing a cascade of ongoing disorders. Furthermore, stock density might explain the poor leg status of cows from urban areas (Solano et al. 2015). In the present study, the farms with higher stock density were located in urban areas with an average of 5.4 animals per herd. In rural areas, stock density was lower with 3.5 cows per herd. Moreover, the type of sheds might influence hock health as well. Additionally, in urban areas, the high density of buildings implies a drastic shortage of farmland, forcing urban farmers to utilise the basement of their houses as cattle stables (Fig. 10.1). Concrete floors, as mostly found in urban areas, increase the predisposition for lameness (Singh et al. 1993) and hock lesions (Norring et al. 2008).

In 73% of the urban farms in this study, cattle "graze" freely in the bustling streets of Bengaluru, implying an increased risk for uncontrolled mating and contribution to genetic variation within herds. The grazing percentage in the selected rural farms was 10% lower than in urban areas.

[1] 0.35–0.45 €/liter at the time of the study.

[2] 0.29–0.31 €/liter at the time of the study.

10.5 Conclusions

Urban dairy farming operates under environmental challenges and faces unfavourable animal husbandry conditions with detrimental impact on animal health and on genetic intra-herd diversity. At the same time, the urban market conditions contribute to intensification processes, stimulating urban farmers to improve dairy cow productivity.

Acknowledgements The authors gratefully acknowledge the kindness, patience and help of the dairy farmers in their fieldwork as well as the logistic and scientific support offered by the partners at the National Institute of Animal Nutrition and Physiology (NIANP) and the financial support provided by the German National Science Foundation, Deutsche Forschungsgemeinschaft (DFG), through grant numbers KO 3520/5-1 and SCHL 587/6-1 and by the Indian Department of Biotechnology, DBT, as part of the Indo-German consortium of DFG Research Unit FOR2432/1.

References

Bates D, Mächler M, Bolker BM, Walker SC (2015) Fitting linear mixed-effects models using lme4. J Stat Softw 67:1–48

Chao B, Dongmei H (2015) The causal relationship between urbanization, economic growth and water use change in provincial China. Sustainability 7:16076–16085

de Vries M, Bokkers EAM, Dijkstra T, van Schaik G, de Boer IJM (2011) Invited review: associations between variables of routine herd data and dairy cattle welfare indicators. J Dairy Sci 94:3213–3228

Hoffmann EM, Jose M, Nölke N, Möckel T (2017) Construction and use of a simple index of urbanization in the rural–urban interface of Bangalore, India. Sustainability 9:2146

Lombard JE, Tucker CB, von Keyserlingk MAG, Weary DM (2010) Associations between cow hygiene, hock injuries, and free stall usage on US dairy farms. J Dairy Sci 93:4668–4676

Norring M, Manninen E, de Passillé AM, Rushen J, Munksgaard L, Saloniemi H (2008) Effects of sand and straw bedding on the lying behavior, cleanliness, and hoof and hock injuries of dairy cows. J Dairy Sci 91:570–576

Office of the Registrar General & Census Commissioner, India (2014) Sample registration system, Baseline Survey 2014. http://www.censusindia.gov.in/vital_statistics/BASELINE%20TABLES07062016.pdf. Accessed on 6 Oct 2019

Oxford Economics (2018) Global cities. Which cities will be leading the global economy in 2035? https://www.oxfordeconomics.com. Accessed on 6 Oct 2019

Rutherford KMD, Langford FM, Jack MC, Sherwood L, Lawrence AB, Haskell MJ (2008) Hock injury prevalence and associated risk factors on organic and nonorganic dairy farms in the United Kingdom. J Dairy Sci 91:2265–2274

Singh SS, Ward WR, Lautenbach K, Murray RD (1993) Behaviour of lame and normal dairy cows in cubicles and in straw yard. Vet Rec 133:204–208

Solano L, Barkema HW, Pajor EA, Mason S, LeBlanc SJ, Zaffino Heyerhoff JC, Nash CGR, Haley DB, Vasseur E, Pellerin D, Rushen J, de Passillé AM and Orsel K (2015) Prevalence of lameness and associated risk factors in Canadian Holstein-Friesian cows housed in freestall barns. J Dairy Sci 98:6978–6991

Part IV
Food Systems as Outcomes and Drivers of Rural-Urban Transitions

Chapter 11
Crop Diversity, Dietary Diversity, and Nutritional Intake: Evidence on Inter-Linkages Across the Rural–Urban Interface of Bengaluru

Veerabhadrappa Bellundagi, K. B. Umesh, B. C. Ashwini, and Stephan von Cramon-Taubadel

Abstract In India, the linkages between agriculture and nutrition are complex and often debated in the policy discourse. Over the past two decades, agricultural production and per capita availability of food grains have increased. In this context, the study has examined the food consumption patterns, the relationship between agricultural production and dietary diversity, and the impact of dietary diversity on nutritional intake. It used a sample of 1275 households in the rural–urban interface, within two transects north and south of Bengaluru. Based on a 14-day food consumption recall, the Simpson Index of Dietary Diversity was used to analyze the dietary diversity and Herfindahl Index was employed to assess the diversity in crop production. The dietary diversity is a good indicator for assessing nutritional adequacy, as it is positively associated with energy and nutrients intake. An estimate of diversity scores revealed that the dietary diversity was marginally higher in northern transect (0.80) compared to the southern transect (0.79). Within the transects, the highest dietary diversity was observed in urban areas, while production diversity was higher in the rural than in the transition area. The findings showed that agricultural production diversity is a major determinant of dietary diversity which in turn has a strong effect on calorie and protein intake. Promoting agricultural diversification will thus help enhancing food and nutrition security in the country.

Keywords Crop diversification · Diets · Household survey · Simpson Index of Dietary Diversity · Herfindahl Index · Nutrition patterns

V. Bellundagi (✉) · K. B. Umesh · B. C. Ashwini
Agricultural Economics, University of Agricultural Sciences Bangalore, Bengaluru 560065, India
e-mail: veeru.b4619@gmail.com

K. B. Umesh
e-mail: umeshuas@gmail.com

S. von Cramon-Taubadel
Agricultural Policy, Agricultural Economics and Rural Development, Georg-August-University Göttingen, 37073 Göttingen, Germany

E. Hoffmann et al. (eds.), *The Rural-Urban Interface*, The Urban Book Series,
https://doi.org/10.1007/978-3-030-79972-4_11

11.1 Introduction

Agriculture is the primary source of livelihood for a sizable portion of the world's rural population. In India, the linkages between agriculture and nutrition are complex and often debated in the policy discourse (Kumar et al. 2016). In most of the developing countries, over the past two decades, crop diversity declined with the spread of green revolution technologies (Jacques and Jacques 2012). A series of innovations in agricultural research and development boosted agricultural production worldwide, mostly in rice (*Oryza sativa* L.), wheat (*Triticum* spp.), and maize (*Zea mays* L.), and per capita availability of food grains has increased, also in India. However, as the majority of farmers in the country's developing regions are subsistence oriented and their diets mainly depend on the crops they harvest, it seems reasonable that diversified agricultural production would lead to a more diverse diet. Thus, farming practice leading to decreased crop diversity and causing nutrition deficiencies was recognized as a major issue in both, subsistence and market-oriented production. For better health and elimination of malnutrition, a diversified and balanced food basket is required (Sangeetha et al. 2013).

Agricultural farm diversification can influence the household's dietary diversity. The crops cultivated by a farming household are direct sources of food (Kavitha et al. 2016; Pandey et al. 2016). On the other hand, increased agricultural productivity raises household income, enabling households to purchase additional food items and achieve dietary diversity and good nutrition (Hawkes and Ruel 2006). Therefore, crop diversification would enhance consumption of a wider variety of food groups and nutritional intake (Chatterjee 2016). However, information on linkages between food production and consumption pattern seems lacking. The present study attempts to connect these variables in a holistic approach and identify a causal link between crop diversity and dietary diversity. It examined the food consumption patterns, the relationship between agricultural production and dietary diversity, and the impact of dietary diversity on nutritional intake.

11.2 Data and Methodology

The study used a stratified random sampling technique for the selection of households from rural, transitional, and urban areas in two transects across the rural–urban interface north and south of Bengaluru (N- and S-transect, respectively). The distinction of rural, transitional, and urban areas was based on the Survey Stratification Index developed by Hoffmann et al. (2017). In the N-transect, 616 households and in the S-transect 659 samples were selected randomly constituting a total sample size of 1275 households. The required data was collected using a structured questionnaire, which also addressed contextual information to allow relating the dietary diversity to other socioeconomic characteristics.

11.2.1 Analytical Techniques Employed

The paper specifically examined the crop diversity, dietary diversity, and the impact of dietary diversity on nutritional intake of the sample households. The quantity of food items consumed by the respondent households was recorded based on a 14-day recall period. Questionnaires were quite exhaustive with almost all types of food items (12 food groups) generally consumed by rural–urban inhabitants in India represented. Further, diversity in the consumption basket of sample households was analyzed in detail using the Simpson Index of Dietary Diversity (SIDD) as indicated in Eq. (11.1).

$$\text{SIDD} = 1 - \sum_{i=1}^{N} P_i^2 \tag{11.1}$$

where P_i is the proportion of the ith food item in total monthly consumption of all food commodities by the members of household. The monthly estimates were subsequently averaged to get the final SIDD estimate for the year under consideration. The Simpson Index ranges between 0 and 1, where the value moves toward 0 in case of complete specialization. Separate scores of SIDD were obtained for households belonging to each village.

The information on area allocated to different crops, allied activities undertaken, and production realized was recorded by the resident investigators from the sample household on a regular basis. The crop diversity was analyzed using Herfindahl Index (HI) as shown in Eq. (11.2).

$$\text{HI} = \sum_{i=1}^{N} P_i^2 \tag{11.2}$$

where P_i represents acreage proportion of the ith crop in total cropped area. The Herfindahl Index takes the value of 1 when there is specialization (only one crop) and approaches 0 when there is diversification and a large number of different crops are grown.

11.3 Results and Discussion

The dietary diversity is a good indicator for assessing nutritional adequacy. The dietary diversity of food is positively associated with energy and nutrients intake (Tarini et al. 1999; Rose et al. 2002; Kant 2004). It is also positively correlated with the three pillars of food security, viz. availability, access, and utilization (Bernal and Lorenzana 2003; Styen et al. 2006; Hillbrunner and Egan 2008). Therefore, estimating household dietary diversity can be an alternative and easier pathway to understand household-level food and nutrition security (Thorne-Lyman et al. 2010; Headey and Ecker, 2013; Taruvinga et al. 2013).

Based on the recommended daily calorie intake, 71.3% of the households were found to be food secure in the N-transect and 67.5% in the S-transect, respectively (Bellundagi et al. 2020a; Chap. 12). The households' dietary pattern as analyzed by the SIDD showed that the dietary diversity, too, was marginally higher in the N-transect compared to the S-transect (Table 11.1). Within the transects, the highest dietary diversity was observed in the urban area (0.83 and 0.82) mainly due to changes in life styles and income of the households leading to diversity in the consumption basket.

Various factors were associated with dietary diversity across the rural–urban interface of Bengaluru (Table 11.2). Per capita income, irrigation, livestock keeping, position in the urban area, and in the N-transect had a positive and significant influence on dietary diversity, while family size had a negative influence on dietary diversity.

The cropping pattern followed in the study area showed that more than 30 crops were grown (see also Chap. 1), with the major area under ragi crop (105 ha and 149 ha in S- and N-transect, respectively). Overall, the gross cropped area and net cropped area were higher in the S-transect (423 ha and 348 ha, respectively) than in

Table 11.1 Simpson Index of Dietary Diversity for sample households across the rural–urban interface of Bengaluru, India

Areas	N-transect	S-transect
Rural	0.78	0.77
Transition	0.79	0.79
Urban	0.83	0.82
Overall	0.80	0.79

Note SIDD = "0" No diversification, SIDD = "1" Complete diversification

Table 11.2 Factors associated with dietary diversity across the rural–urban interface north of Bengaluru, India, using fractional probit model

Variables	Coefficients	z value
Per capita income (Rs./month) [INR]	0.000031*	4.73
Irrigation (D1)	0.179*	2.84
Livestock (D2)	0.0903***	1.68
Family size (no.)	−0.039*	−3.47
Urban (D3)	0.231**	1.99
Transition (D4)	0.077	1.45
Transect (D5)	0.091***	1.67
Constant	0.893*	6.63
Log pseudolikelihood value	−134.52	

Note D1 to D5 = dummies; *Significant at 1%, **Significant at 5%, ***Significant at 10%

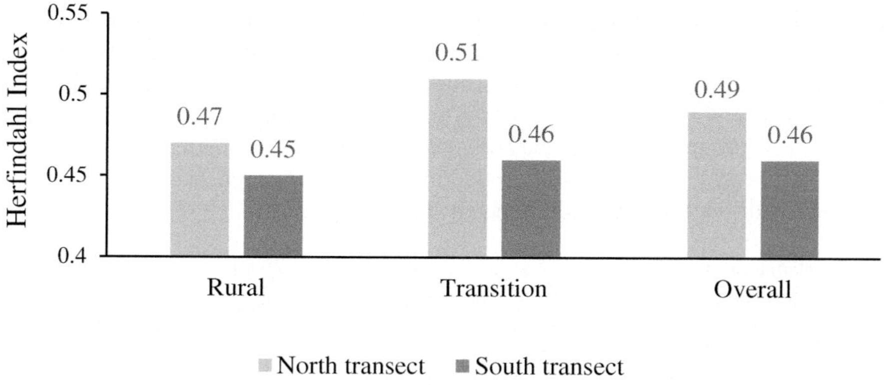

Fig. 11.1 Crop diversification by farmers across the rural–urban interface of Bengaluru, India. Note: HI = 0 Total diversification, HI = 1 Total specialization

the N-transect (415 ha and 340 ha). The production diversity index (HI) was 0.49 in the N-transect and 0.46 in the S-transect. In eastern India, in a region characterized by high crop diversification (Kumar et al. 2012), the diversification index was 0.44, which compares well to the present study. Adding to the higher production diversity index in the S-transect, the agriculture sector was shown to be the major source of employment for rural households (33%) in the agriculture-dominated areas of Bengaluru (Bellundagi et al. 2020b).

Along the urbanization gradient, differences were marginal in the S-transect. In the N-transect, however, production diversity was highest in rural area, whereas specialization was highest in the transition area (Fig. 11.1). Contrary to this, Udaykumar et al. (Chap. 4) had observed increasing diversification toward the urban region among a subsample of farmers that had availed agricultural credit. Furthermore, an increasing specialization toward vegetable farming in transitional and urban regions as shown by Nandini et al. (Chap. 3) would likely imply higher diversity at the species level (Chap. 1). However, the studies by Udaykumar et al. and Nandini et al. (Chaps. 3 and 4) analyzed much smaller household samples with a specific focus and may thus not be representative.

11.4 Conclusions

The linkages between agriculture and nutrition are complex, but need to be pursued to shape the agenda of agricultural development conducive for enhancing nutrition among the farming households. Although our results point toward an association between agricultural production portfolios with diet quality, further in-depth research is needed to explicitly identify the precise nutritional benefits. By quantifying both parameters in a large household sample with suitable diversity indices, this study has prepared the grounds for a better understanding of the relationships between production diversity and dietary diversity in the rural–urban interface of Bengaluru. The results highlighted that agricultural production diversity may be a key determinant of dietary diversity. The dietary diversity in turn has a strong effect on calorie intake. Agricultural programs and policies oriented toward reducing under-nutrition should promote diversity in agricultural production rather than emphasizing on increasing production through focusing on selected staple crops.

Acknowledgements This paper is a part of Indo-German Collaborative research project titled "The Rural-Urban Interface of Bangalore—A Space of Transition in Agriculture, Economics and Society"—Sub-project "Food Insecurity at Different Stages of Urbanisation." The Department of Biotechnology (DBT), GOI, New Delhi, India (Sanction order: BT/IN/German/DFG/14/BVCR/2016), is duly acknowledged for the financial support for this project.

References

Bellundagi V, Umesh KB, Ashwini BC (2020a) An economic assessment of the food security of households in the rural-urban interface of Bangalore. Agric Econ Res Rev 33:39–47

Bellundagi V, Umesh KB, Ashwini BC (2020b) Is food insecurity exists among the households in rural-urban interface of Bangalore? An economic analysis. J Food Agric Environ 18(3 & 4):27–35

Bernal JR, Lorenzana PA (2003) Dietary diversity and associated factors among beneficiaries of 77 child care centers: Central regional, Venezuela. Arch Latinoam Nutr 53(1):52–58

Chatterjee T (2016) Does farm level diversification improve household dietary diversity? Evidence from Rural India. Paper presented at the 90th annual conference of the agricultural economics society, Warwick University, Coventry, 4–6 April 2016

Hawkes C, Ruel M (2006) Understanding the links between agriculture and health. 2020 Focus 13. Brief 1. IFPRI, Washington, DC, USA

Headey D, Ecker O (2013) Rethinking the measurement of food security: from first principles to best practice. Food Secur 5(3):327–343

Hillbrunner C, Egan R (2008) Seasonality, household food security and nutritional status in Dinajpur, Bangladesh. Food Nutr Bull 29(3):221–231

Hoffmann E, Jose M, Nölke N et al (2017) Construction and use of a simple index of urbanization in the rural–urban interface of Bangalore, India. Sustainability 9(11):1–21

Jacques PJ, Jacques RC (2012) Monocropping cultures into ruin: the loss of food varieties and cultural diversity. Sustainability 4(11):2974–2997

Kant AK (2004) Dietary patterns and health outcomes. J Am Dietetic Assoc 104(4):615–635

Kavitha K, Soumitra P, Padmaja R (2016) Understanding the linkages between crop diversity and household dietary diversity in the semi-arid regions of India. Agric Econ Res Rev 29:129–137

Kumar A, Kumar P, Sharma AN (2012) Crop diversification in Eastern India: status and determinants. Indian J Agric Econ 67(4):600–617

Kumar A, Saroj S, Singh KP et al (2016) Agricultural diversity, dietary diversity and nutritional intake: An evidence on inter-linkages from village level studies in Eastern India. Agric Econ Res Rev 29:15–29

Pandey VL, Dev SM, Jayachandran U (2016) Impact of agricultural interventions on the nutritional status in South Asia: a review. Food Policy 62:28–40

Rose D, Meershoek S, Ismael C et al (2002) Evaluation of a rapid field tool for assessing household diet quality in Mozambique. Food Nutr Bull 23:181–189

Sangeetha V, Sharma JP, Burman RR et al (2013) Food security vs nutritional Security: need for multi-sectoral convergence. Int J Agric Food Sci Technol 4(6):621–626
Styen PN, Nel HJ, Nantel G et al (2006) Food variety and dietary diversity scores in children: are they good indicators of dietary adequacy? Public Health Nutr 9(5):644–650
Tarini A, Bakari S, Delisle H (1999) The overall nutritional quality of the diet is reflected in the growth of Nigerian children. Cahiers D'études Et De Recherches Francophones / Santé 9(1):23–31
Taruvinga A, Muchenje V, Mushunje A (2013) Determinants of rural household dietary diversity: the case of Amatole and Nyandeni districts, South Africa. Int J Dev Sci 2(4):2233–2247
Thorne-Lyman AL, Valpiani N, Sun K et al (2010) Household dietary diversity and food expenditures are closely linked in rural Bangladesh: increasing the risk of malnutrition due to the financial crisis. J Nutr 140(1):182S-188S

Chapter 12
Food Security and Factors Influencing Food and Calorie Intake Across the Rural–Urban Interface of Bengaluru

B. C. Ashwini, K. B. Umesh, Veerabhadrappa Bellundagi, P. S. Prasannakumar, and Stephan von Cramon-Taubadel

Abstract India is experiencing a puzzling contradiction of food and nutritional insecurity as measured by calorie intake, despite its impressive economic growth. This study quantifies the food and calorie intake to analyse the status of food security and factors influencing it along the rural–urban interface of Bengaluru. It uses primary data from 1275 households in two research transects. The Food Security Index was employed to assess the status of food security. The dietary profile of the households showed that most of the calorie intake was obtained from the consumption of cereals followed by edible oils, milk and milk products, but intake of calories varied significantly along the rural–urban gradient. The factors influencing food and calorie intake were analysed by multiple linear regression. Landholding and per capita income were significantly and positively influencing the status of food security, whereas family size showed a negative influence. Education, family size and landholding had a significant influence on food and calorie intake. Urban households were more food secure, followed by transitional and rural households. In order to increase food security of the rural and transitional households, the existing food security programmes need to be strengthened further.

B. C. Ashwini (✉) · K. B. Umesh · V. Bellundagi · P. S. Prasannakumar
Agricultural Economics, University of Agricultural Sciences Bangalore, GKVK, Bengaluru 560065, India
e-mail: ashwinismile813@gmail.com

K. B. Umesh
e-mail: umeshuas@gmail.com

V. Bellundagi
e-mail: veeru.b4619@gmail.com

P. S. Prasannakumar
e-mail: shiprasannamk415@gmail.com

S. von Cramon-Taubadel
Agricultural Policy, Agricultural Economics and Rural Development, Georg-August-University Göttingen, 37073 Göttingen, Germany
e-mail: scramon@gwdg.de

E. Hoffmann et al. (eds.), *The Rural-Urban Interface*, The Urban Book Series,
https://doi.org/10.1007/978-3-030-79972-4_12

Keywords Nutrition patterns · Household survey · Food Security Index · Multiple linear regression analysis

12.1 Introduction

The green revolution in India had a remarkable impact on agricultural productivity and food production. As a result, India had a surplus production and consumption of cereals since 1960s. Even though India is more or less self-sufficient in cereals, it is deficit in pulses and oilseeds (Praduman et al. 1995; Goyal and Singh 2002; Acharya 2009). Due to changes in consumption patterns, the demand for fruits, vegetables, dairy, meat, poultry and fish has increased. There is a need to have proper crop diversification plans to meet these demands (Lorenzo and Luca 2014; Prakash 2018).

Food insecurity is about people not knowing if and when their next meal will come and not being able to afford the food they want to eat. Food security is ensured if people can always buy the basic food they are accustomed to and the amount of calorie consumed per day is equal to the recommended calorie intake (Kakwani and Son 2016). Along with the status of food and calorie intake, it is important to determine the factors influencing food and calorie intake, to understand actual reasons for the observed status of food secure or insecure households.

12.2 Data and Methodology

12.2.1 Sampling Frame

This study was set up to investigate the status of food security and factors influencing the same in the rural–urban interface of Bengaluru, across which a northern transect (N-transect) and a southern research transect (S-transect) were defined. Each transect was further divided into rural, transitional and urban areas following a remote sensing-based stratification (Hoffmann et al. 2017; Chap. 1). The sample frame consisted of 1275 households, of which, after removing partially filled schedules and outliers, we used 1046 households for the analysis. Those comprised 510 samples from the S-transect (110 urban, 211 transitional and 189 rural households) and 536 samples from the N-transect (73 urban, 150 transitional and 313 rural households) of Bengaluru.

12.2.2 Analytical Techniques

12.2.2.1 Food Security Index

The study employed the Food Security Index (FSI) to assess the food security status of the households. A food security line was determined and used to classify households into either being food secure or food insecure depending on which side of the line they fell. The household caloric acquisition method according to Hoddinott (2000) is the number of calories or nutrients available for consumption by household members over a defined period of time. The household calorie intake was obtained from the household per day consumption. The quantity of every food item consumed by the household in a day was converted into its calorie content. This was achieved by multiplying all respective food items (weight in kilograms) by the corresponding food energy content. This was further converted into per capita calorie by dividing the estimated total household calorie intake by the adjusted household size in adult equivalents (consumption unit). A household was regarded as food secure if daily per capita calorie intake was up to, or above the recommended value, and as food insecure if it was below the threshold. In a similar way, as used by Ibrahim and Bello (2008), the Food Security Index was calculated as follows (Eq. 12.1):

$$\text{FSI} = \frac{\text{Daily per capita calorie intake}}{\text{Recommended daily per capita calorie intake}} \tag{12.1}$$

12.2.2.2 Multiple Linear Regression Analysis

To determine the factors influencing food and calorie intake, a multiple linear regression analysis was carried out. The function for factors influencing food security (Eq. 12.2) was as follows:

$$Y = a_0 + a_1X_1 + a_2X_2 + a_3X_3 + a_4X_4 + a_5X_5 + a_6D_1 + a_7D_2 + a_8D_3 + \mu \tag{12.2}$$

where

Y = Food security index (FSI)

X_1 = Family size (no.)

X_2 = Per capita income (INR/month)

X_3 = Agriculture and allied (man-days/year)

X_4 = Off-farm (man-days/year)

X_5 = Non-farm (man-days/year)

D_1 = Urban (1 = if place of residence is urban, otherwise '0')

D_2 = Transition (1 = if place of residence is transition, otherwise '0')

D_3 = North transect (1 = if place of residence is in north transect, otherwise 0')

a_i (i = 1, ..., 8) = Regression coefficients

μ = Random disturbance term

For determining the factors influencing calorie intake, the function (Eq. 12.3) had the form:

$$\begin{aligned} Y = {} & a_0 + a_1 X_1 + a_2 X_2 + a_3 D1 + a_4 D_2 + a_5 D_3 + a_6 X_3 \\ & + a_7 X_4 + a_8 D_4 + a_9 X_5 + e \end{aligned} \tag{12.3}$$

where

Y = Calorie intake (kcal/capita/day)

X_1 = Age (years)

X_2 = Education (no. of years)

D_1 = Gender (1 = male, otherwise female)

D_2 = Urban (1 = if place of residence is urban, otherwise '0')

D_3 = Transition (1 = if place of residence is transition, otherwise '0')

X_3 = Family size (no.)

X_4 = Landholding (ha)

D_4 = Irrigation (1 = access to irrigation, otherwise '0')

X_5 = Per capita income (INR/month)

a_i (i = 1, ..., 9) = Regression coefficients

e = Random disturbance term.

12.3 Results and Discussion

12.3.1 Status of Food Security in the Rural–Urban Interface of Bengaluru

The share of food secure households in the rural–urban interface of Bengaluru (Table 12.1) reached 75.2% in the S-transect, while in the N-transect, it was 78.2% implying that the majority of the households were food secure. Along the S-transect, food secure households amounted to 63.1%, 73.2% and 83.4% in rural, transitional and

Table 12.1 Status of food security in the rural–urban interface of Bengaluru, India, as expressed by the percentage of food secure households (HH)

South transect	**Rural**	**Transitional**	**Urban**	**Total**
(n)	(189)	(211)	(110)	(510)
Food secure HH	63.12	73.24	83.36	75.16
Food insecure HH	36.88	26.76	16.64	25.94
Total	100.00	100.00	100.00	100.00
North transect	**Rural**	**Transitional**	**Urban**	**Total**
(n)	(313)	(150)	(73)	(563)
Food secure HH	63.23	75.42	85.38	78.23
Food insecure HH	36.77	24.58	14.62	22.87
Total	100.00	100.00	100.00	100.00
Pooled	**Rural**	**Transitional**	**Urban**	**Total**
(n)	(502)	(316)	(183)	(1046)
Food secure HH	63.18	74.33	84.37	76.69
Food insecure HH	36.82	25.67	15.63	24.41
Total	100.00	100.00	100.00	100.00

urban areas, respectively. Within the N-transect, food security status was 63.2% in the rural area, followed by 75.4% in the transitional and 85.4% in the urban area. Food security was thus higher in the urban area of both transects than in rural and transitional areas. Both, per capita income and per capita consumption of calories were relatively higher in urban than in rural and transitional areas, leading to higher food security in the urban area. Similar results were reported by Sabur et al. (1997), indicating that the difference between the low income and the rich households in terms of nutritional intake was wider in the urban than in the rural area.

12.3.2 Factors Influencing Food Security

To examine the impact of family size, land holding, per capita income, position in the urban and transitional region, and in the N-transect on FSI score, the multiple linear regression model was used (Table 12.2). The estimates of the determinants of food security revealed that variables included in the model explained up to 38% of the variation in food security, and the calculated F value was statistically significant. Among the dependent variables included in the model, employment from agriculture and non-farm sources, per capita income and position in the urban area were the significant variables positively influencing the status of food security, while family size showed a negative influence on food security. The study conducted by Mannaf and Uddin (2012) revealed that the per capita food availability declined as family size increased due to population growth. Hence, large family size is more likely related

Table 12.2 Factors influencing food security across the rural–urban interface of Bengaluru, India, using multiple linear regression analysis: dependent variable = Food Security Index (FSI), n = 1046

Variables	Coefficients	t value
Family size (no.)	−0.072*	0.000
Per capita income (Rs./month)	0.0017**	0.049
Employment generation (man-days/year)		
(a) Agriculture and allied	0.0003*	0.003
(b) Off-farm	0.0002	0.051
(c) Non-farm	0.0003*	0.000
Urban (D1)	0.088**	0.045
Transition (D2)	0.087	0.061
North transect (D3)	0.054	0.084
Constant	1.12*	0.000
R^2 value	0.38	
F value	15.06*	

Note *significant at 1%, **significant at 5%; dummies: Urban (D1): 1 = urban, otherwise '0', Transition (D2): 1 = transition, otherwise '0', North transect (D3): 1 = N-transect, otherwise '0'

to food insecurity in a household. The study by Abu and Soom (2016) found that income of the household head, rural household size and farm size had a positive relationship with household food security, while the age of the household head and urban household size had a negative impact on household food security.

12.3.3 Factors Influencing Calorie Intake

Our results indicate that education level and family size had negative and significant influence on calorie intake (Table 12.3). This is similar to the results of Srivastava and Chand (2017), who estimated that family size had an adverse effect on per capita intake of energy, while age and education of the respondents had positive and significant effects on food and nutritional status of the households. Our study identified that consumption of cereals was more than the recommended intake throughout both, the N- and S-transects. However, educated people were more aware that consuming highly calorie-based food items (especially cereals) leads to obesity and other health-related problems; hence, education had a negative relationship with per capita calorie intake. Landholding size, on the other hand, had a positive and significant influence on the calorie intake. On average, if the sample respondent stayed in the urban area, he/she will have an additional calorie intake of 174 kcal/capita/month compared to individuals staying in rural and transitional area. These results are in line with the study conducted by Kumar et al. (2016). The data presented here further served as a

Table 12.3 Factors influencing calorie intake across the rural–urban interface of Bengaluru, India, using regression analysis: dependent variable = calorie intake (kcal/capita/month), $n = 1046$

Variables	Coefficients	t value
Age (years)	−3.896	−1.120
Education (no. of years)	−17.085***	−1.850
Gender (D1)	142.196	1.44
Urban (D2)	174.116***	1.87
Transition (D3)	25.182	0.36
Family size (no.)	−60.92*	−3.45
Landholding (ha)	80.44	1.81
Irrigation (D4)	51.017	0.38
Per capita income (INR./month)	0.001	1.26
Constant	1020*	0.00
R^2 value	0.374	
F value	8.22*	

Note *Significant at 1%, ***significant at 10%; dummies: Gender (D1): 0 = female, 1 = male; Urban (D2): 1 = urban, otherwise ‘0’; Transition (D3): 1 = transition, otherwise ‘0

basis for a more detailed analysis of the composition of diets and the contribution to calorie intake by different food items (Bellundagi et al. 2020).

12.4 Conclusions

This study illustrated that food insecurity still remains an issue in rural, transitional and urban areas of both, the N- and S-transect of Bengaluru, in spite of many central and state government food security programmes implemented in the country. Hence, in order to increase and strengthen the status of food security among rural, transitional and urban households, the existing food security programmes need to be strengthened further. This also necessitates interventions to educate households to change/modify their existing pattern of food and calorie intake to reduce the gap between recommended dietary allowances and actual calorie intake. Regression results indicated that education, family size and landholding size significantly influenced calorie intake.

Acknowledgements This paper is a part of Indo-German Collaborative Research project titled ‘The Rural–Urban Interface of Bengaluru—A Space of Transition in Agriculture, Economics and Society’—Sub-project ‘Food Insecurity at Different Stages of Urbanisation’. The Department of Biotechnology (DBT), GOI, New Delhi, India (Sanction order: BT/IN/German/DFG /14/BVCR/2016 dated 16th November 2016) is duly acknowledged for the financial support for this project.

References

Abu GA, Soom A (2016) Analysis of factors affecting food security in rural and urban farming households of Benue state, Nigeria. Int J Food Agric Econ 4(1):55–68

Acharya SS (2009) Food security and Indian agriculture: policies, production performance and marketing environment. Agric Econ Res Rev 22:1–19

Bellundagi V, Umesh KB, Ashwini BC (2020) Is food insecurity exists among the households in rural-urban interface of Bangalore? An economic analysis. J Food Agric Environ 18(3 & 4):27–35

Deaton A, Drèze J (2009) Food and nutrition in India: facts and interpretations. Econ Pol Wkly 44(7):42–65

Goyal SK, Singh JP (2002) Demand versus supply of food grains in India: implications to food security. Paper presented at 13th International Farm Management Congress, Wageningen, The Netherlands, July 7–12, 2002

Hoddinott J (2000) Operationalizing household security and development strategies. International Food Policy Research Institute, Technical Guideline No. 1, Washington, DC, USA

Hoffmann E, Jose M, Nölke N et al (2017) Construction and use of a simple index of urbanization in the rural–urban interface of Bangalore India. Sustainability 9(11):1–21

Ibrahim H, Bello M (2008) Food security and resource allocation among farming household in North Central Nigeria. Department of Agricultural Economics and Extension, Nassarawa State University, Keffi, Nigeria

Kakwani N, Son HH (2016) Social welfare functions and development: Measurement and policy applications. Palgrave Macmillan, London

Kumar A, Saroja S, Singh RKP et al (2016) Agricultural diversity, dietary diversity and nutritional intake: an evidence on inter-linkages from village level studies in Eastern India. Agric Econ Res Rev 29:15–29

Lorenzo P, Luca T (2014) Crop diversification, dietary diversity and agricultural income: empirical evidence from eight developing countries. Canad J Dev Stud 35(2):211–227

Mannaf M, Uddin TM (2012) Socio-economic factors influencing food security status of maize growing households in selected areas of Bogra district. Bangladesh J Agric Econ 35(1 & 2):177–187

Praduman K, Mark R, Peter H (1995) Cereals prospects in India to 2020: implications for policy. International Food Policy Research Institute (IFPRI) 2020 Brief 23, June 1995

Prakash CD (2018) Diversification of agriculture: a review. J Econ Soc Dev 14(1):46–59

Sabur SA, Talukdar RK, Kabir MH (1997) Consumption pattern of food commodities in an area of Mymensing. Bangladesh J Agric Econ 20(2):93–106

Srivastava SK, Chand R (2017) Tracking transition in calorie-intake among Indian households: insights and policy implications. Agric Econ Res Rev 30(1):23–35

Chapter 13
Household and Individual Nutrition in the Rural–Urban Interface of Bengaluru

Anjali Purushotham

Abstract Using primary socioeconomic survey data, we studied the nutrient adequacy and nutritional status of households and individuals in the rural–urban interface of Bengaluru (India), respectively. We found that on average the sample households consumed recommended levels of macro-nutrients but lacked recommended consumption of important micro-nutrients. While the nutritional status of children indicated significant undernourishment, overnutrition was a rising concern among women. This shows the complex situation of the coexistence of multiple forms of malnutrition in the rural–urban interface of Bengaluru.

Keywords Consumption patterns · Household survey · Diets · Macro- and micro-nutrients · Anthropometric measurements · Nutritional status

13.1 Introduction

Despite the rapid growth in its economy and declining poverty rates, India continues to be one of the largest contributors to global hunger (Global Hunger Index 2020). This is reflected in the sluggish improvements in the anthropometric outcomes of children (NFHS-4 2015–16). While there is a need for a stronger commitment from the government to tackle undernutrition, the concerns of overweight and obesity are rising rapidly. Over the last decade, the rates of overweight/obesity among men and women in India have doubled (NFHS-4 2015–16). The latest data (CNNS 2019) show that there is a growing risk of non-communicable diseases among children aged 5–9 years and adolescents aged 10–19 years. These statistics suggest that India is facing a complex situation of slow improvements in undernutrition and rising concerns of overnutrition simultaneously. Literature has emphasized the need for a better understanding of nutrition transition, the sociocultural context of food acquisition, and vulnerable sub-populations, to design effective interventions for sound

A. Purushotham (✉)
Agricultural Policy, Agricultural Economics and Rural Development, Georg-August-University Göttingen, 37073 Göttingen, Germany
e-mail: anjali.katiganere-purushotham@agr.uni-goettingen.de

E. Hoffmann et al. (eds.), *The Rural-Urban Interface*, The Urban Book Series,
https://doi.org/10.1007/978-3-030-79972-4_13

and sustainable economic growth in India (Meenakshi 2016; Shankar et al. 2017; Siddiqui et al. 2017; Aizawa 2019).

Given the fast paced growth and urbanisation in India, there are many areas where the boundaries between rural and urban are fuzzy. We refer to such peri-urban zones as the rural–urban interface. The rural–urban interface is a dynamic environment and expected to bring about change in food and non-food consumption patterns, health outcomes, lifestyle, and living conditions. People living in such a dynamic rural–urban interface are subjected to unique combinations of key drivers of nutrition such as income growth, access to own production but also modern marketing channels and sources of food, and lifestyle changes. Studying the status of nutrition in this special setting where urbanisation is unfolding at an especially rapid pace can help us understand the trends in urbanisation and nutrition transition. Therefore, this study descriptively presents the household nutrient adequacy and inadequacy for three macro-nutrients (calories, protein, and fat) and three micro-nutrients (vitamin A, iron, and zinc), and the nutritional status of children and women.

13.2 Materials and Methods

We used household-level data from a primary socioeconomic survey conducted in the two research transects of the FOR2432 project (Chap. 1). Following the stratified random sampling design, we analyzed 1126 households in the rural–urban interface of northern and southern Bengaluru.[1] The household decision maker and the family caregiver were interviewed to collect household socioeconomic and food consumption data, respectively. In addition, anthropometric measurements were conducted on children and women in these sample households.

Households' macro- and micro-nutrient adequacy and inadequacy are calculated in three steps. First, the household members' demographic information (such as age and gender) was used to calculate the recommended quantities of each nutrient (considered in the study) that should theoretically be consumed by our sample households.[2] In the second step, the 14-day recall food consumption data were used to calculate the total quantities of each nutrient actually consumed by the same households, following the nutrient conversion factors from the Indian Food Composition Table (Longvah et al. 2017). In the third step, the total quantity of each nutrient consumed by the household was divided by the total quantity recommended for that particular household to get the nutrient adequacy ratios. The households for which the adequacy ratio for a nutrient is more than one are considered as nutrient adequate and

[1] While the survey consists of 1275 households, the household member's demographic and food consumption data were available only for 1126 households. Therefore, our study is based on these 1126 households for whom complete sociodemographic and food consumption data were available.

[2] Indian Council of Medical Research (2009) provides the recommended dietary allowances (RDAs) of nutrients for the different demographic populations in India. The RDAs for each of the nutrients considered in the study for each member of our sample households are added together to calculate the respective nutrient requirements for our sample households.

the households for which the nutrient adequacy ratio is less than one are considered as nutrient inadequate with respect to that particular nutrient.

The nutritional status of children and women was calculated by z-scores and Body Mass Index (BMI), respectively. Detailed information regarding the sampling design, household data collection, household nutrient consumption, and individual nutritional status can be found in Purushotham et al. (2018).

The households were sampled using a stratified sampling design. That is, using the Survey Stratification Index (SSI), each transect was divided into six strata from which the villages were randomly selected. Stratum 1 was the most urban, and stratum 6 the most rural (Hoffmann et al. 2017; Chap. 1). These six strata were grouped into three regions, i.e., the sample villages in strata 1 and 2 represent the urban, strata 5 and 6 the rural, and strata 3 and 4 the transitional region. The following sections of the paper address the percentage of households that consume less than the recommended macro- and micro-nutrients, and the nutritional status of children and women in rural, urban, and transitional areas of the rural–urban interface of Bengaluru. While the assessment of food security presented by Ashwini et al. (Chap. 12) was based on estimated household calorie availability only, the present study considers a broad range of macro- and micro-nutrients in the diet consumed. Furthermore, this study presents the nutritional status of children and women in the sample households.

13.3 Results and Discussion

13.3.1 Household Macro- and Micro-nutrient Consumption

On average, the sample households in Bengaluru met their macro-nutrient requirements. However, a closer look at the estimates shows that 28%, 25%, and 7% of the households did not consume recommended quantities of calories, protein, and fat, respectively. Furthermore, a fair proportion of them did not consume recommended quantities of micro-nutrients. That is, 70%, 49%, and 29% of our sample households consumed less than the recommended quantities of vitamin A, iron, and zinc. Public-health policy in India is mainly concentrated on improving the access to major staple food items to meet the protein-energy requirements, and less importance is given on meeting the micronutrient requirements of the population (Drèze 2012; Pingali et al. 2017). This is also reflected in our study area, where households on average consume sufficient quantities of macro-nutrients but severely lack in the consumption of micro-nutrients (Table 13.1). In our study, there was no significant difference in the prevalence of macro-nutrient adequate and inadequate households between the three regions. However, there were significant differences in the prevalence of micronutrient adequate and inadequate households at different stages of urbanisation. While vitamin A inadequacy was most critical altogether, and became severe toward transitional and rural areas, inadequacy of iron and zinc seemed to be more problematic in urban areas of Bengaluru.

Table 13.1 Nutrient adequate and inadequate households in the rural–urban interface of Bengaluru, India

Nutrient	Factor	Urban (%)	Transition (%)	Rural (%)
Calorie	Adequate	69	73	72
	Inadequate	31	27	28
Protein	Adequate	75	76	73
	Inadequate	25	24	27
Fat	Adequate	94	95	91
	Inadequate	6	5	9
Vitamin A	Adequate	43	34	22
	Inadequate	57	66	77
Iron	Adequate	42	53	53
	Inadequate	58	47	47
Zinc	Adequate	63	72	74
	Inadequate	37	28	26
N	1126			

13.3.2 Nutritional Status of Children

The nutritional status of 276 children aged 0–5 years was assessed through the weight-for-age (WAZ), height-for-age (HAZ), and weight-for-height (WHZ) z-scores (World Health Organization 2006). Among our sample children, 24% were underweight (WAZ $\leq$ 2 SD), 36% were stunted (HAZ $\leq$ 2 SD), and 16% were wasted (WHZ $\leq$ 2 SD). Underweight and wasting indicate short-term changes in the nutritional status of a child, which is often associated with acute starvation and/or severe disease. Stunting is associated with inadequate nutrition over the long term. The undernutrition rates for our sample children in Bengaluru are slightly lower than the national-level statistics.[3] As discussed before, the rural–urban interface of Bengaluru is a rapidly urbanising region where the socioeconomic status of the households can be slightly higher than the national average. This may be the reason for lower undernutrition rates than the national-level statistics. However, the rate of stunting among children is almost equal to the national-level statistics indicating the severity of this problem even in a rapidly urbanising region such as Bengaluru. Our data show that the prevalence of underweight was lower in urban than in rural and transitional regions (Fig. 13.1). There was not much difference in the prevalence of stunting among the three regions. The urban region had a higher prevalence of wasting among children than the other two regions. However, none of these differences were statistically significant.

[3] At the national level, there are 36%, 38%, and 21% of children below 5 years who are underweight, stunted, and wasted, respectively (NFHS-4 2015–16).

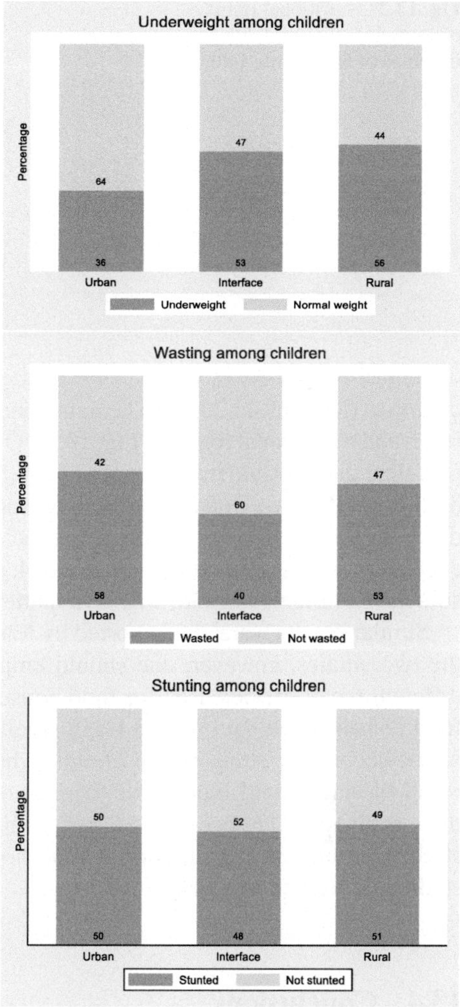

Fig. 13.1 Nutritional status of children in the rural–urban interface of Bengaluru, India

13.3.3 Nutritional Status of Women

The nutritional status of 1438 women was measured by calculating their BMI. In our sample, 16% of the women were thin (BMI < 18.5 kg/m^2), 48% had normal weight (BMI $\geq$ 18.5 to < 25 kg/m^2), and 36% were overweight or obese (BMI $\geq$ 25 kg/m^2). Compared to the national-level statistics, the rate of thinness was lower in our sample (22% at the national level, NFHS-4 2015–16). However, the percentage of women who were overweight or obese in the rural–urban interface of Bengaluru was higher than the overweight/obesity prevalence at the national level (21% NFHS-4 2015–16). This is a rising concern because studies show that non-communicable diseases associated with overweight and obesity are expected to rise (see also Chap. 14). These occur at lower rates of fat intake and overnutrition in India compared to developed

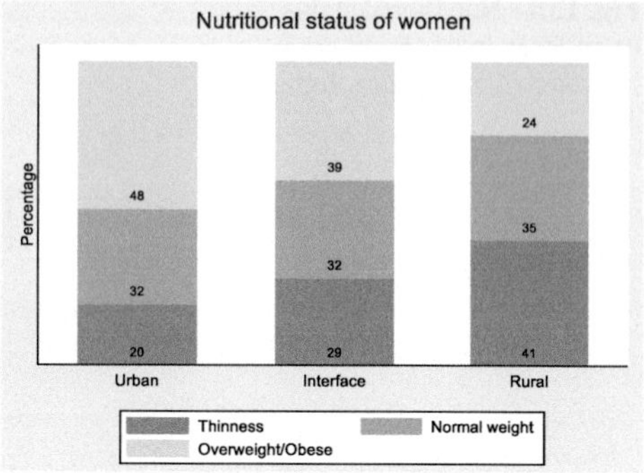

Fig. 13.2 Nutritional status of women in the rural–urban interface of Bengaluru, India

countries (Meenakshi 2016). The nutritional status of women varied between urban, rural, and transitional regions (Fig. 13.2). Overweight/obesity was significantly more prevalent in urban and transitional regions, while the prevalence of thinness was very high in rural areas. Understanding how such differences in the nutritional status at different stages of urbanisation are associated with the changes in dietary preferences (Chap. 15), agricultural (Chaps. 3, 4, and 5), and/or non-agricultural livelihood strategies of the household, will be a matter of further research.

Similar results were also reported by Yatnatti et al. (Chap. 14). Before comparing the two studies, however, one should emphasize the scale of observation and the different recall periods used for food consumption data. The current study used the food consumption data for a 14-day recall period and considered the entire range of socioeconomic strata of households, whereas Yatnatti et al. (Chap. 14) used 24-h recall of actual food intake and focused on middle-class households. In addition, there is also a difference in the target sample of the two studies. The present paper studied the nutrient consumption at household level, whereas Yatnatti et al. (Chap. 14) studied nutrient intake among women.

13.4 Conclusions

The rural–urban interface of Bengaluru is a region where the divide between rural and urban is fuzzy and constitutes a transitional area exposed to very dynamic transformations. The food consumption and thereby the nutritional status of people are influenced by the unique combination of different factors and thus show complex patterns. This can be observed in the nutrient consumption of households and the nutrition status of children and women in our sample from the rural–urban interface. Most of the sampled households in Bengaluru acquire recommended norms of macronutrients; however, they often lack in the consumption of recommended quantities of the micro-nutrients studied. There is a significant undernutrition prevalence among children and overnutrition prevalence among women living in these households. This shows the complex situation of concurrent micronutrient deficiency, undernutrition, and overnutrition. Further exploration of the factors determining each of these aspects and interactions between them helps to improve the understanding of the nutrition situation in the rural–urban interface of Indian megacities.

Acknowledgements The authors gratefully acknowledge the financial support provided by the German Research Foundation, DFG, through grant number 279374797 as part of the Research Unit 2432/1. We would also like to thank our partners at the University of Agricultural Sciences Bangalore, and in particular Prof. K. B. Umesh for coordination and logistic support.

References

Aizawa T (2019) Transition of the BMI distribution in India: evidence from a distributional decomposition analysis. J Bioecon 21(1):3–36

CNNS (2019) Nutrition Survey finds rising diabetes risk in children. Comprehensive National Nutrition Survey (2016–18). In The Economic Times, 2019. Available online at https://economictimes.indiatimes.com/news/politics-and-nation/nutrition-survey-finds-rising-diabetes-risk-in-children/articleshow/71487940.cms. Checked on 8 Sept 2019

Drèze J (2012) Food and nutrition. In: Basu K, Maertens A (eds) The new Oxford companion to economics in India, 3rd edn. Oxford University Press, New Delhi

Global Hunger Index (2020) Global Hunger Index 2020: India: Available online at https://www.globalhungerindex.org/india.html

Hoffmann E, Jose M, Nölke N et al (2017) Construction and use of a simple index of urbanisation in the Rural-Urban Interface of Bangalore India. Sustainability 9(11):1–21

Indian Council of Medical Research (2009) Nutrient requirements and recommended dietary allowances for Indians. Available online at http://citeseerx.ist.psu.edu/viewdoc/download?doi=10.1.1.627.5927&rep=rep1&type=pdf. Checked on 27 May 2020

Longvah T, Ananthan R, Bhaskarachary K et al (2017) Indian food composition tables. Hyderabad

Meenakshi JV (2016) Trends and patterns in the triple burden of malnutrition in India. Agric Econ 47(S1):115–134. https://doi.org/10.1111/agec.12304

NFHS-4 (2015–16) National family health survey. India Fact Sheet. International Institute for Population Sciences, Mumbai

Pingali P, Mittra B, Rahman A (2017) The bumpy road from food to nutrition security–slow evolution of India's food policy. Glob Food Sec 15:77–84

Purushotham A, Mittal N, Aswini BC et al (2018) Dietary diversity and children anthropometric outcomes: a quantile regression analysis. Göttingen: Georg-August-Universität Göttingen, Courant Research Centre - Poverty, Equity and Growth (CRC-PEG) (Discussion Papers UR. https://www.econstor.eu/handle/10419/181999, 259)

Shankar B, Agrawal S, Beaudreault AR et al (2017) Dietary and nutritional change in India: implications for strategies, policies, and interventions. Ann N Y Acad Sci 1395(1):49–59

Siddiqui MZ, Donato R, Jumrani J (2017) Looking past the Indian calorie debate: what is happening to nutrition transition in India. J Dev Stud 1–20

World Health Organization (2006) WHO child growth standards: length/height-for-age, weight-for-age, weight-for-length, weight-for-height and body mass index-for-age: methods and development

Chapter 14
Nutritional Status of Women Across Rural–Urban Interface of Bengaluru

K. Geetha, D. Vijayalakshmi, Shilpa Yatnatti, Christoph Dittrich, and Deepa S. Patil

Abstract Urbanisation influences the nutritional status as a result of fundamental changes in food consumption patterns and lifestyle. The assessment of women's nutritional status is of greater public health significance as they are more likely to suffer from malnutrition than men. A total of 300 middle-income households were selected from three areas representing different urbanisation stages. Anthropometric measurements, nutrient intake and prevalence of non-communicable diseases were studied among women in selected households. The findings revealed that under-nutrition was more prevalent among rural women (13.6%), while 34.9% of women in the rural–urban transition zone were obese. The highest proportion of women with increased waist-to-hip ratio was observed among the urban dwellers (58.7%). The per cent adequacy intake of fat (262.3%) and calcium (190.0%) was above the recommended dietary allowances, whereas iron (83.0%) and fibre (79.6%) did not meet the requirements. The share of women suffering from non-communicable diseases decreased from urban (16.0%) towards transitional (11.2%) and rural (10.2%) regions. Dietary intake tending towards high-fat and low-fibre foods is directly linked to the health and nutritional status of women in the rural–urban interface of Bengaluru.

Keywords Consumption patterns · Middle class · Household interviews · Diets · Anthropometric measurements · Gender · Health status

---

K. Geetha · D. Vijayalakshmi · S. Yatnatti (✉) · D. S. Patil
Food Science and Nutrition, University of Agricultural Sciences Bangalore, Bengaluru 560065, India
e-mail: shilpayatnatti@gmail.com

D. S. Patil
e-mail: deepapatil07@gmail.com

C. Dittrich
Human Geography, Georg-August-University Göttingen, 37077 Göttingen, Germany
e-mail: christoph.dittrich@geo.uni-goettingen.de

E. Hoffmann et al. (eds.), *The Rural-Urban Interface*, The Urban Book Series,
https://doi.org/10.1007/978-3-030-79972-4_14

14.1 Introduction

A diet adequate to maintain the composition of nutrients in the body and normal metabolic integrity is the requirement for human health. How well these requirements are met is represented by the nutritional status (OMICS 2014). To assess the nutritional status, anthropometric parameters are valuable instruments. Growing urbanisation impacts the nutritional status of individuals by influencing socio-demographic, health, and food consumption patterns.

Urbanisation is the increasing share of a nation's population living in urban areas. Most urbanisation is the result of net rural to urban migration and involves changes in occupation patterns, lifestyles, family structures, and value systems (Satterthwaite et al. 2010). Changes in the health and nutritional status result from fundamental changes in food consumption patterns and lifestyle. Women are more likely to suffer from malnutrition than men, for reasons including women's reproductive biology, sociocultural traditions, and disparities in household work patterns (Joshi et al. 2016). Assessment of women's nutritional status is thus of greater public health significance. Hence, the present investigation focused on the nutritional status of women in regions at different levels of urbanisation, along a presumed gradient across the rural–urban interface of Bengaluru.

14.2 Materials and Methods

Sample selection. Based on purposive random sampling, 300 middle-income households were selected for the assessment of their food consumption pattern and nutritional status. They belonged to two research transects north and south of Bengaluru, each subdivided into rural, transitional and urban regions based on a survey stratification index (SSI; Hoffmann et al. 2017; Chap. 1). In each transect, 50 households were selected from rural, transitional and urban villages, respectively.

Anthropometric measurements. To assess the nutritional status, the following anthropometric measurements were recorded: height, weight, waist circumference and hip circumference. Indices such as body mass index (BMI) and waist–hip ratio (WHR) were calculated based on the anthropometric measurements.

Nutrient intake. Dietary intake of food and nutrients was assessed for 50% of the female respondents through 24 h recall method for three days in one week, whereby two consecutive days and one weekend day were considered. Women were asked to recall the type and quantity of foods consumed for breakfast, lunch, evening tea, dinner and any other meal of the previous day (other than feasting and fasting day). Quantities were estimated using standardised cups, vessels, paper discs and rubber balls. Data on intake of food items such as cereals, pulses, fats, vegetables, fruits, milk and milk products and sugar were evaluated. Based on the quantity of foods consumed per day, nutrient intake was calculated using tables of nutritive value of Indian foods (Longvah et al. 2017). These figures were compared against the recommended dietary allowance (RDA) for women as per the revised requirements suggested by the Indian Council of Medical Research (ICMR 2010) to provide a measure of adequacy or inadequacy of nutrient consumption.

Non-communicable diseases. Along with the interviews and measurements, the women in the selected households were interrogated about non-communicable diseases mainly such as diabetes mellitus (DM) and hypertension (HTN).

All collected information was consolidated and analysed for significance with the chi-square test to draw interpretations towards the study objectives.

14.3 Results

14.3.1 Anthropometric Measurements

Body Mass Index. The study revealed differences in the nutritional status of middle-class women in the regions along the rural–urban gradient (Table 14.1). Under-nutrition was more among rural women (13.5%), while about 20.0% of urban women were overweight. It was interesting to note that 34.87% of women were obese (grade I) in the transition zone, which was more compared to rural (27.1%) and urban (30.0%). These differences in the distribution of BMI were statistically significant with $\chi^2 = 29.54$.

On average, 53.7% of women across the rural–urban interface of Bengaluru had increased WHR compared to the standards. Among the study areas, the highest proportion of increased WHR was observed among urban women (58.7%).

Table 14.1 Nutritional status of women across the rural–urban interface of Bengaluru, India

	Rural		Transitional		Urban		
BMI class	(n)	%	(n)	%	(n)	%	χ^2 test
Under-nutrition (<18.5)	24	13.56	10	6.58	8	5.33	29.54**
Normal (18.5–22.9)	61	34.47	50	32.90	44	29.33	
Overweight (23–24.9)	29	16.38	27	17.76	30	20.0	
Pre-obese (25–29.9)	48	27.12	53	34.87	45	30.0	
Obese I (>30)	15	8.47	12	7.89	23	15.34	
Total	177	100.0	152	100.0	150	100.0	

**Significance at 5%

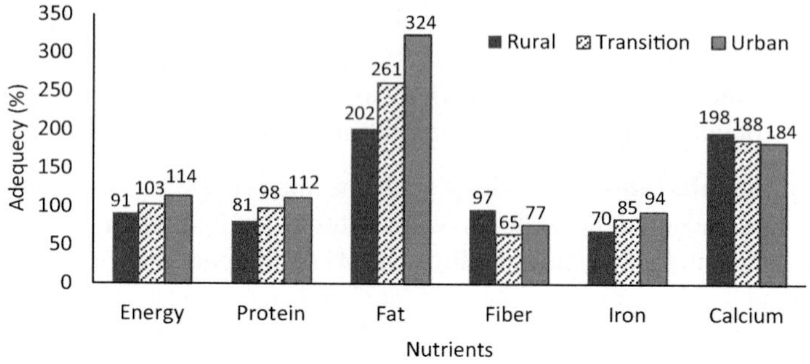

Fig. 14.1 Nutrient intake among women across the rural–urban interface of Bengaluru, India

14.3.2 Nutrient Intake

Nutrient intake of women was compared with RDA and expressed as per cent adequacy (Fig. 14.1). In the rural area, energy (91%) and protein (81%) intake almost met RDA, but iron intake was only 70% adequate. Energy (103%) and protein (98%) intake were adequate in the transitional region, but exceeded recommended values in the urban region. Fat intake was almost two- to threefold higher among women in all the study areas and highest among urban women (324%). Fibre intake was comparatively low among transitional (65%) and urban (77%) women. Iron (Fe) intake was close to adequacy in urban (94%), but deficient in transitional (85%) and rural (70%) regions. Calcium (Ca) intake was found to be almost double the requirement among women across the rural–urban interface of Bengaluru.

14.3.3 Prevalence of Non-communicable Diseases

A total of 16% of women were suffering from non-communicable diseases in urban, followed by transition (11.2%) and rural (10.2%) regions (Fig. 14.2). Among non-communicable diseases, prevalence of diabetes was highest among urban women (9.3%). However, about 5.1% of the rural women were having hypertension, which was observed to be 2.6% and 4.7% among women in the transitional and urban areas, respectively.

14.4 Discussion

According to the World Health Organization (WHO), "malnutrition, in all its forms, includes under-nutrition, inadequate vitamins or minerals, overweight, obesity,

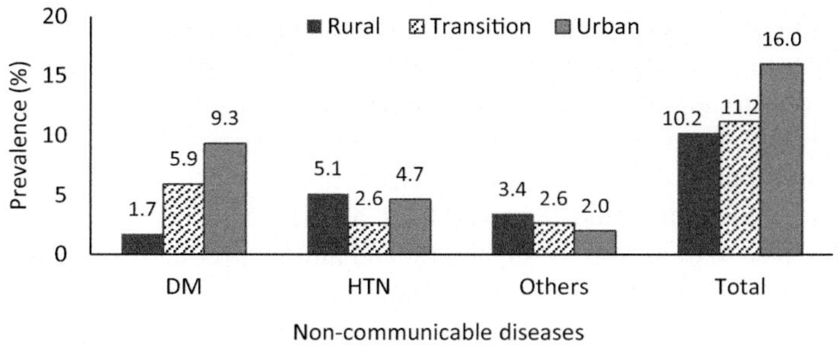

Fig. 14.2 Incidence of non-communicable diseases among women across rural–urban interface of Bangalore, India. DM: Diabetes mellitus, HTN: hypertension

and resulting diet-related non-communicable diseases" (WHO 2020). Urbanisation involves changes in occupation patterns, lifestyles, family structures and value systems. These changes are reflected on dietary practices and levels of physical activity which have an impact on the nutritional status of individuals and households. A fundamental cornerstone of any individual's health is "adequate nutrition", and it is especially critical for women, because inadequate nutrition affects not only women's health, but also the health of children. In this regard, understanding the magnitude of malnutrition along a gradient across the rural–urban interface of Bengaluru helps to derive the influence of urbanisation on women's nutritional status.

The BMI is an effective tool to measure adult nutritional status, both of individuals and of communities, and helps to draw conclusions on prevalence of related health issues. The BMI assessment indicated the prevalence of under-nutrition was higher among rural women, which is in line with findings reported by Vir and Malik (2015). In their study of the nutrition situation of women in India, the prevalence of under-nutrition was higher, and the difference between rural areas (40.6%) and urban regions (25.0%) was larger. According to the National Family Health Survey (NFHS) fact sheet of the Karnataka state, conducted in 2015–16, malnutrition among rural women was 24.3%, as compared to 16.2% in urban areas. Various reasons, including sociocultural practices, beliefs and taboos related to food consumption, may affect the women's dietary pattern leading to malnutrition. The lower reported magnitude for under-nutrition in the present study may be attributed to the shift from under-nutrition to over-nutrition as a result of growing urbanisation influence of Bengaluru megacity on surrounding rural localities and to its focus on middle-class households.

The surprising fact that 34.87% of women were obese (grade I) in the rapidly changing rural–urban transition area gives a clear indication of transition in nutritional status of women especially living in the interface of rural and urban. Over-nutrition of women is an emerging problem in India with serious implications on adult onset of non-communicable diseases. The rate of over-nutrition at national level is 12.6%, with Delhi, Kerala, Punjab and Tamil Nadu having over 25.0% of over-nourished women (Vir and Malik 2015). The findings of the present study commemorate NFHS data for Karnataka (2015–16), which indicated a prevalence of obesity among urban women of 31.8%. The changing lifestyle pattern attributing to variations in physical activity and food consumption is the leading cause for increasing obesity among urban women. Increased WHR an indicator of abdominal obesity was also more among urban women, correlating with the highest prevalence of obesity in urban.

Nutrition transition is the term used for a shift from traditional diets high in cereals and fibre to diets high in sugars, fat and animal products, often referred to as "Western diets" (Chap. 15). The nutrient intake among women across the rural–urban interface of Bengaluru revealed two- to threefold higher intake of fat than adequate. Supportive to this, Vatsala et al. (2017) reported fat intake ranging between 33 and 35 g/day (RDA = 20 g/day) among selected rural women in south India. A gradual shift was observed for fat intake from rural to urban indicating that this is one of the urbanisation-related factors in nutrition transition. Increased intake of low-cost, processed convenience foods may be attributed to the high fat intake. A major staple across all areas of the rural–urban gradient of Bengaluru is finger millet (ragi). It is used in different food preparations, such as ragi ball, roti or porridge, and is an important source for dietary calcium; hence, irrespective of the area about twofold higher intake of calcium than the RDA was observed. However, there are chances that calculated values may be higher than the analytical values and prevalence of phytates restricts bioavailability of calcium (Harinarayan and Ramalakshmi 2015). An increasing trend was observed for the energy intake and an inversely decreasing trend for fibre intake among women. Consumption of more refined and processed foods, associated with urbanisation, increases energy intake and decreases the fibre intake.

Epidemiological studies have reported India as one of several countries that face high prevalence of pathologies related to the metabolic syndrome, notably type 2 diabetes, dyslipidemia, obesity and cardiovascular disorders. Research indicates that these diseases are also impacted by dietary habits with a strong association with malnutrition (Sébastia et al. 2013). Prevalence of non-communicable diseases was highest among urban women and may be correlated to BMI which is one of the important risk factors. According to Htet et al. (2016), the prevalence of metabolic risk factors, such as overweight, obesity, hypertension, hypercholesterolemia, hypertriglyceridemia and diabetes, was highest among urban residents, while the behavioural risk factors such as smoking, alcohol consumption and low fruit and vegetable intake were more commonly reported among rural people. Similar observations were made in the present study.

14.5 Conclusions

Urbanisation leads to shifts in dietary intake towards high-fat and low-fibre foods. This nutrition transition is directly linked to the prevalence of a dual burden of malnutrition with under and over-nutrition at the same time. The strong association of over-nutrition with the incidence of non-communicable diseases adversely affects the health status of women across rural–urban interface of Bengaluru. In this regard, it is necessary to understand corresponding risk factors which may be identified and addressed with proper channels in order to protect and promote women's health status.

Acknowledgements This work was part of the Indo-German research project "The Rural-Urban Interface of Bengaluru - A Space of Transitions in Agriculture, Economics, and Society". We are thankful to the Department of Biotechnology (DBT), Govt. of India, for financial support.

References

Harinarayan CV, Ramalakshmi T (2015) Patterns of dietary calcium intake in South Indian rural, urban and metropolitan city subjects. J Clin Sci Res 4(2):143–148

Hoffmann EM, Jose M, Nölke N et al (2017) Construction and use of a simple index of urbanization in the rural–urban interface of Bangalore, India. Sustainability 9(11):1–21

Htet AS, Bjertness MB, Sherpa LY et al (2016) Urban-rural differences in the prevalence of non-communicable diseases risk factors among 25–74 years old citizens in Yangon Region, Myanmar: a cross sectional study. BMC Public Health 16(1):1–12

ICMR (2010) Nutrient requirements and recommended dietary allowances for Indians. National Institute of Nutrition, Hyderabad, India, pp 160–162

Joshi P, Sharma N, Singh BK (2016) Health and nutrition issues of rural women in India. J Agric Eng Food Technol 3(2):87–90

Longvah T, Ananthan R, Bhaskarachary K et al (2017) Nutritive value of Indian foods, National Institute of Nutrition (ICMR). Hyderabad, India, pp 1–168

NFHS-4 (2015–16) State fact sheet Karnataka. Ministry of Health and Family Welfare, Government of India, pp 1–4

OMICS (2014) Nutritional status research articles available from: https://www.omicsonline.org/nutrition-disorder-and-therapy/nutritional-status-research-articles.php

Satterthwaite D, McGranahan G, Tacoli C (2010) Urbanisation and its implications for food and farming. Philos Trans R Soc Biol Sci 365(1554):2809–2820

Sébastia S, Balagopal P, Misra R (2013) Diet related diseases: Issues and solutions to nutrition transition and food programme policies in India. Research gate. Available on: https://www.researchgate.net/publiction/236143159

Vatsala L, Prakash J, Prabhavathi SN (2017) Food security and nutritional status of women selected from a rural area in South India. J Food Nutr Popul Health 1(2):1–8

Vir S, Malik M (2015) Nutrition situation of women in India: current status, implications on child under-nutrition and challenges ahead. Statist Appl 13(1 & 2):71–84

WHO—World Health Organisation (2020) Malnutrition. https://www.who.int/news-room/fact-sheets/detail/malnutrition

Chapter 15
"Homemade Food is Always Healthy": The Ambivalence of a Traditionalisation of Diets in Bengaluru

Mirka Erler, Neda Yousefian, and Christoph Dittrich

Abstract In India and elsewhere, traditional diets are praised for their ostensibly positive impacts on consumer's health and environmental sustainability. Promoting their perpetuation, or in some cases rediscovery, is supposed to prevent people's alienation from locally sourced food. However, "traditional diet" is also a vague concept that does not particularly address the aims mentioned above. In this contribution, we will discuss the possible affordances of traditional diets for health and sustainability, while at the same time questioning the application of the term tradition. Recent developments in the rural–urban interface of the South Indian megacity Bengaluru are presented as a case study. We argue that some aspects of Bengaluru's traditional diets, for example the high consumption of whole grains such as finger millet, are indeed conducive to health and sustainability. However, the concept of "traditional diet" is prone to fallacies about the capacities of some traditional dishes, while the hegemony over the definition of what is traditional provides the basis to exclude certain groups based on their food practices.

Keywords Consumption patterns · Middle class · Household interviews · Dietary preferences · Food transition

M. Erler (✉) · N. Yousefian · C. Dittrich
Human Geography, Georg-August-University Göttingen, 37077 Göttingen, Germany
e-mail: mirka.erler@thuenen.de

N. Yousefian
e-mail: neda.yousefian@uni-goettingen.de

C. Dittrich
e-mail: christoph.dittrich@geo.uni-goettingen.de

M. Erler
Institute of Market Analysis, Thünen Institute, 38116 Braunschweig, Germany

E. Hoffmann et al. (eds.), *The Rural-Urban Interface*, The Urban Book Series,
https://doi.org/10.1007/978-3-030-79972-4_15

15.1 The Traditionalisation of Diets

Modern Indian middle-class diets are increasingly becoming a problem for human health and for the environment.[1] An increased consumption of high caloric food, partly induced by a globalization of diets, leads to obesity and other diet-related diseases (Pingali and Khwaja 2004; Law 2019; Chap. 14). The increased consumption of animal products, in particular milk and poultry, although relatively moderate in comparison with other Asian countries, increases the environmental footprint of human nutrition (Tilman and Clark 2014). Awareness about these problems exists in India, with people particularly concerned with the health implications of modern diets. Our research in the rural–urban interface of the south Indian mega-city Bengaluru revealed an apparent return of middle-class households to traditional diets, as one possible way to alleviate the developments mentioned above (Erler and Dittrich 2017). In this contribution, we refer to this phenomenon as the traditionalisation of diets.

Our understanding of the traditionalisation of diets can best be described as a hybrid term between a more critical notion of traditionalisation which is predominantly applied in social sciences and a more practical and positivist notion of traditional diets which is predominantly applied in food sciences and economics. The social sciences regard traditionalisation as a process in which alleged practices from the past are applied in the presence (Adkins 1999; Darakchi 2018). This process involves a certain degree of othering of what is conceived as non-traditional practices (Stahl 2018). In food sciences and economics, the term traditional diet usually refers to some characteristics which have been dietary features in a particular place and time period of the past. These features are then either regarded as problematic for, or as conducive to, health (Pingali and Khwaja 2004; Cárdenas-Castro et al. 2019; Law 2019; Rossi 2019). While we acknowledge that the conclusions drawn about the features of diets in these contributions can be described as accurate, they are not paying attention to the ambiguity of the term tradition. In this contribution, we therefore regard the traditionalisation of diets as a process in which spatially and temporally confined features of diets are evaluated, recommended or applied regarding their affordances for modern diets. However, these evaluations often disregard questions of hegemony over definition and social othering.

In this contribution, we will discuss the up- and downsides of a traditionalisation of diets. We will summarize findings from our research in the city of Bengaluru and the surrounding rural–urban interface, where we observe that a traditionalisation of diets is taking place. Our aims are to show the relevance of the traditionalisation of diets as a theoretical concept, to point out the concept's ambiguity, and to derive research questions to explore this topic further in Bengaluru and elsewhere.

---

[1] In this article, we are primarily dealing with over-nutrition and its consequences, which is increasingly becoming a problem even for lower classes in India. However, it should not be left unsaid that concomitantly hunger and under-nutrition are persisting in India today. This coexistence of over- and under-nutrition is usually referred to as the double burden of malnutrition.

15.1.1 The Up- and Downsides of a Traditionalisation of Diets

Contributions discussing the benefits of traditional diets often refer to one particular food item or dish. The benefits of those foods are often a high content of fresh vegetables and spices (Cárdenas-Castro et al. 2019), locally produced food in general, locally produced cereals, pulses (Pingali and Khwaja 2004; Law 2019), fish, and a low content of processed meat (Rossi 2019). Furthermore, many traditional dishes are cooked from scratch in the households (Pingali and Khwaja 2004), which gives the people a high degree of control over what is added to their food and perpetuates the knowledge about preparing food within the population. Although most publications have a specific idea on what they regard as traditional, none of them unequivocally defines what a traditional diet is, thus, they leave the term up for interpretation by the reader.

Despite not providing a clear definition of what a "traditional" diet is, some of these publications provide another unclear term of what it is *not*, namely "Western". By contrasting the traditional food items and dishes to Western diets, these contributions go on to generalize the benefits of usually only one particular food discussed, to traditional food in general. It is often implied that the discussed foods are not only healthy because they are traditional, but because they contain particularly healthy ingredients or components, as described in the previous paragraph.

Pointing at "the Western diet" as the source of problems in contemporary diets in the Global South has indeed some justification. While many foods items with high fat and sugar contents, such as burgers or chocolate bars, are Western in their origin, they are imported to the Global South by an industrial food system rather than by a joined effort of "the West", like the term Western diet might suggest. Thus, in our study, we suggest avoidance of the term because it obscures the actual promoter of unhealthy and often highly processed foods, namely food industries. Following the logic of traditional vs. Western food, a company could replace an unhealthy Western food product, with an unhealthy "traditional" food product, and while this would correspond to the demand to increase the consumption of traditional food, the health outcome would remain the same.

Another point of contention when discussing the traditionalisation of diets is the question of whose food tradition is being upheld. While common staple grains and vegetables are not exclusively consumed by one social group, there are distinct dietary practices that fall along religious and class lines, specifically a non-vegetarian diet or the consumption of beef. The latter has become a rallying point for fundamentalist Hindus who use politics, and in some cases physical violence, to stop this specific traditional food practice (Sathyamala 2019). Such efforts to homogenize and dominate what constitutes a traditional diet have so far failed to diminish the actual diversity of traditional diets; however, the politics behind these efforts persist in promoting a narrative of what traditional is, and perhaps even more clearly, what it is not, regardless of these religious and class distinctions.

Finally, we must raise the issue of who is responsible for preparing the traditional foods. The expectation that women, who in previous generations engaged in the time-consuming labor of preparing traditional foods from scratch for their families, continue to perform this duty, while in many cases maintaining employment outside of the household promotes outdated gender stereotypes. In acknowledging a new generation of middle-class working women, Pingali and Khwaja (2004) associate the decline of traditional meals with women joining the workforce and the subsequent consequences for the family diet. This statement may seem like an objective observation; however, in light of the movement for traditional foods, it also speaks to upholding oppressive gender roles for the sake of the family's health and well-being.

15.1.2 Traditionalisation of Diets in Bengaluru

We base our assessment on the traditionalisation of diets in Bengaluru on two main findings. First, households in Bengaluru's rural–urban interface communicated a dichotomous classification of healthy and unhealthy foods based on whether a food was a traditional food or not. Second, the organic food movement in Bengaluru, represented by Alternative Food Networks (AFN) and their customers, highlight the value of local and traditional foods for peoples' diets and the environment.

According to households in the rural–urban interface, a dish was particularly healthy when it was typically consumed in rural areas, made at home, and Indian. Food which was typically consumed in urban areas, eaten out or foreign, specifically Western, was regarded to be unhealthy. Dishes and food items which fell under the first category were ragi balls (a lump made of soaked finger millet flour), sambar (a light soup of lentils and vegetables), and local fruit and vegetables, which were allegedly more commonly consumed in rural areas. Dishes and food items which fell under the second category were, for instance, pani puri (a fried street food popular in India), burgers, and instant noodles. At first glance, these categories appear quite adequate to assess the health impact of a dish or a food. However, we also argue that the categories are neither explicit nor accurate regarding the health impact of food.

The most prominent example for the promotion of a traditional food by Bengaluru's organic movement is the case of millets. While finger millet and jowar are cultivated and consumed in and around Bengaluru, the organic movement is promoting a variety of many different millets to be rediscovered by consumers. Millets are promoted to be healthy, environmentally friendly, and inherent to the Indian subcontinent (Erler et al. 2020). A second example is the promotion of so-called Desi or A2 milk. This milk is sourced allegedly only from native Indian cattle races (such as Sahiwal or Gir, though not specified in the interviews) and has a different beta-casein structure than conventional milk (Truswell 2005). The arguments for the promotion of this milk range from the preservation of local cow races to the compatibility of the genes of these cows with those of the Indian population (Erler and Dittrich 2020). The organic movement's argumentation with regard to these examples is highly ambiguous. While it is certainly important to promote the consumption of foods that are beneficial to both human health and the environment, capitalizing on patriotic sentiments of consumers will lead to a commodification of these products. This might change the meaning of millets and A2 milk from popular foods to exclusive commodities.

In order to derive conclusions with regard to exclusionary mechanisms mentioned in the papers presented above, we must acknowledge major limitations in the fact that we mainly spoke to people who identify with the superior group when referring to tradition. In other words, we hardly spoke to those people who might be excluded by the process of traditionalisation.

15.2 Conclusions

In this contribution, we discussed the up- and downsides of a traditionalisation of diets. While those diets which are discussed as traditional do indeed contain a lot of favorable ingredients, such as a high vegetable content, the term is often unprecise and should therefore not be utilized as a dietary recommendation. With regard to our research in Bengaluru's rural–urban interface, we summarized that a traditionalisation of diets currently manifests itself in several ways. Traditionalisation of diets supposedly offers a way for people to assess what food is healthy; in this case, the re-introduction of local grain varieties could be a way to support a more environmentally friendly, less intensive form of agriculture. However, we argue that these movements open doors for misconceptions and social exclusion. Therefore, we suggest that the promotion of beneficial foods, labeled as traditional, should be done by sticking to the factual advantages of the foods, such as particular health or environmental benefits, rather than by blurring these advantages behind equivocal terms like *traditional*.

We call for further research on how the traditionalisation of diets is perceived by different social groups, in order to assess if and how social exclusion through discourses on diets is perceived by them.

Acknowledgements The authors gratefully acknowledge the financial support provided by the German Research Foundation, DFG, through grant number DI 709/2-1 as part of the Research Unit 2432/1-2. We would also like to thank the University of Agricultural Sciences Bangalore, as our Indian partner university, particularly Prof. Vijayalakshmi D. and her staff. Their advice and support have been an important asset for our research in Bengaluru.

References

Adkins L (1999) Community and economy: a retraditionalisation of gender? Theory Cult Soc 16(1):119–139

Cárdenas-Castro AP, Perales-Vázquez GdC, La Rosa LA de et al (2019) Sauces: an undiscovered healthy complement in Mexican cuisine. Int J Gastronomy Food Sci 17(2019):1–11

Darakchi S (2018) Gender, religion, and identity: Modernization of gender roles among the Bulgarian Muslims (Pomaks). Women's Stud Int Forum 70:1–8

Erler M, Dittrich C (2017) Nahrungs und Ernährungsverhalten urbaner Mittelschichten in Bangalore. Geogr Rundsch 12:26–32

Erler M, Dittrich C (2020) Middle class, tradition and the Desi-Realm—discourses of alternative food networks in Bengaluru, India. Sustainability 12:2741

Erler M, Keck M, Dittrich C (2020) The changing meaning of millets: organic shops and distinctive consumption practices in Bengaluru, India. J Consumer Cult 1–19. https://doi.org/10.1177/1469540520902508

Law C (2019) Unintended consequence of trade on regional dietary patterns in rural India. World Dev 113:277–293

Pingali P, Khwaja Y (2004) Globalisation of Indian diets and the transformation of food supply systems (No. 854-2016-56199)

Rossi L (2019) Promotion of the traditional Greek diet in children to enable healthy diets in a nutrition-sensitive way. Public Health Nutr 22:2700–2702

Sathyamala C (2019) Meat-eating in India: whose food, whose politics, and whose rights? Policy Futures Educ 17(7):878–891

Stahl G (2018) Narratives in reconstituting, reaffirming, and (Re)traditionalizing identities. Men Masculinities 21(5):709–728

Tilman D, Clark M (2014) Global diets link environmental sustainability and human health. Nature 515(7528):518–522

Truswell AS (2005) The A2 milk case: a critical review. Eur J Clin Nutr 59:623–631

Part V
Biodiversity and Ecosystem Functions Under Rural-Urban Transitions

Chapter 16
Tree Species Diversity in the Southern Transect Across the Rural–Urban Interface of Bengaluru

B. N. Divakara, C. U. Nikhitha, M. A. Mahmud, Nils Nölke, and V. P. Tewari

Abstract Urbanisation is a complex process, differently defined in different contexts and disciplines. However, a general feature of urbanisation is that it comes along with changing environmental landscapes by accommodating large numbers of migrants and industries in densely built-up areas. This affects environmental and ecological components of the landscape. Trees have many ecological and environmental functions, as they provide habitat and protection for fauna and flora, and support spiritual, cultural and recreational activities of people. Our study addresses the composition, density and diversity of trees in the urban, transitional and surrounding rural region South of Bengaluru, India. WorldView-3 satellite imagery was used to randomly select 25 one-hectare plots with a high share of built-up area to characterize the tree species composition. A total of 82 tree species were observed in these plots, out of which 21 species were found in all the three domains. Diversity indices such as Shannon Index, Simpson Index, Pielou's Evenness, and Margalef's Richness indicated that tree species richness decreased towards the rural domain though the species composition varied from plot to plot. The Shannon and Simpson Index decreased towards the transitional and rural domain. The species were more or less evenly distributed in all domains.

Keywords Vegetation assessment · Tree species diversity · Diversity indices · Importance Value Index · Biomass index

---

B. N. Divakara (✉) · C. U. Nikhitha · V. P. Tewari
Genetics and Tree Improvement, Institute of Wood Science and Technology, Bengaluru 560003, India
e-mail: bndsira@gmail.com

V. P. Tewari
e-mail: tewarivp@gmail.com

M. A. Mahmud · N. Nölke
Forest Inventory and Remote Sensing, Georg-August-University Göttingen, 37077 Göttingen, Germany
e-mail: Nils.Noelke@forst.uni-goettingen.de

E. Hoffmann et al. (eds.), *The Rural-Urban Interface*, The Urban Book Series,
https://doi.org/10.1007/978-3-030-79972-4_16

16.1 Introduction

Urban forestry deals with the management of trees in urban areas to provide and enhance environmental and ecological services. Green areas and, in particular, trees play a crucial role to support urban biodiversity and ecology (Jim and Chen 2009, Chaps. 17 and 18), mitigating greenhouse gas emissions, reducing urban flood, and modifying local climate by reducing the urban heat island effects. In addition, trees in urban areas provide aesthetic benefits and enhance functional diversity. They also contribute towards economic benefits in that they may attract tourism and increase property value and investments. In the process of urbanisation, industrialisation, and other anthropogenic activities, urban green spaces and green cover are at high risk of experiencing loss of species and decline in area due to increasing population and demand for urban expansion (Paul and Nagendra 2015). Montgomery (2008) predicted that about two-third of the global population will reside in cities by 2050. In India alone, there are currently 46 cities with more than one million inhabitants, the largest being the mega cities of Mumbai, Delhi, Kolkata, Chennai, and Bengaluru (Census of India 2011). In this regard, it holds in particular also for India that the green spaces in urban areas must be a primary concern in the process of urbanisation (Colding et al. 2006).

For many decades Bengaluru had a reputation of "Garden City of India" with huge lush trees and green spaces (Basavaraja 1984). The vegetation in the city had a high share of exotic species with high species richness and diversity when compared with peripheral and surrounding rural regions (Sudhira and Nagendra 2013). However, more recently Bengaluru experienced intensive developmental activities, including construction of multi-storied buildings and road widening, which lead to felling of many old huge trees. That has raised concerns of an increasing urban heat island effect. A recent study on Bengaluru suggested that better data and information on tree cover and its characteristics like species composition, size and age structure are essential for a more effective management and protection of street trees and of biodiversity (Nagendra and Gopal 2010). This study presents new data on tree species distribution and species composition in and around the city.

16.2 Materials and Methods

The present study focused on Southern Bengaluru and assessed trees in 25 plots, each of 100 m × 100 m (1 ha) in size, laid out in the transect defined by the Indian-German research consortium of FOR2432 (Fig. 16.1). The division of the transect into urban, transition, and the rural categories was taken on the basis of Awuah et al. (2018). Briefly, the transect was first divided into squares of 3 km^2 north–south direction, and the proportion of built-up pixels was determined by a cross-classification approach based on a WorldView-3 satellite image. Squares with more than 50% built-up pixels were regarded as urban, whereas areas with 10–50% and 0–10% built-up pixels were defined as transitional and rural domains, respectively. Within those domains, the 1 ha

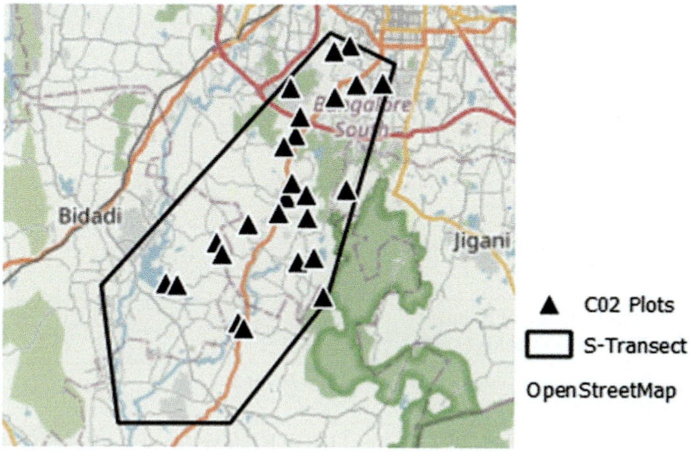

Fig. 16.1 Location of the 25 observation plots (C02 Plots) surveyed along the Southern transect of Bengaluru, India

plots were delineated, and plots with more than 50% impervious surface were defined as built-up layer. Selection of the field plots followed a stratified random sampling. Our 25 field plots were all from the "built-up" layer as our main interest lies in urbanised areas with green spaces such as street trees, trees inside private premises, or field margins.

In all field plots the tree cover was assessed as follows: crowns were delineated, species identified, height and diameter measured (diameter at breast height for trees >10 cm) and vitality scored. These variables were recorded for each tree within each plot. If tree crowns of neighbouring trees formed larger crown patches that could not be disentangled into individual tree crowns, the whole patch was delineated and numbers of trees recorded together with their species.

16.2.1 Indices for Diversity, Biomass, and Importance

Diversity indices quantify the variables within a biological community. In fact, "diversity" is a measure of dispersion for nominal variables like "variance". Variance is a measure of dispersion for metric variables. Indices of biodiversity are an important tool to characterise structure and composition of a biological community and they provide an idea of rarity or commonness of species within a community (Bhat et al. 2016). In the present study, diversity indices such as Shannon–Wiener diversity index (Shannon and Wiener 1963), Simpson's index of dominance (Simpson 1949), Margalef's richness index (Ludwig and Reynolds 1988), and Pielou's evenness index (Pielou 1975) were used (Table 16.1).

The Shannon–Wiener diversity index (H') gives the characterization of species diversity in the vegetative population. The values extend from zero to one. The

Table 16.1 Indices of diversity used in this study

Sl. No	Diversity index	Formula
1	Shannon–Wiener diversity index Shannon and Wiener (1963)	$H' = -\sum_{i=1}^{s} P_i(\ln P_i)$
2	Simpson's index of diversity Simpson (1949)	$D = 1 - \sum_{i=1}^{s}[n_i/N]^2$
3	Simpson's index of dominance Simpson (1949)	$Cd = \sum_{i=1}^{s}[n_i/N]^2$
4	Margalef's index of richness Ludwig and Reynolds (1988)	$R = \frac{S-1}{\ln N}$
5	Pielou's evenness index Pielou (1975)	$e = H'/\ln S$
6	Importance Value Index Tripathi et al. (2004)	IVI = Relative density + Relative frequency + Relative dominance

Where : H'—Shannon's index of diversity, $P_i - n_i / N$ = Proportion of total sample belonging to the *i*th species, *D*—Number of individuals of the species '*i*', *N*—Total number of individuals in the plot, *Cd*—Concentration of dominance, *R*—Margalef Richness index, *S*—Total number of species in a community, *n*—Total number of individuals observed

value increases as both species richness and the evenness of the community increase (Stohlgren 2007). Simpson's dominance index (Cd) considers the number of individual species present as well as the relative abundance (D) of each species. Species richness (R) is a simple count of the number of different species in a given community. The Evenness (e) refers to the distribution pattern of the individuals between the species in the community. Importance Value Index (IVI) was calculated by summing up relative frequency, relative density, and relative dominance values in a particular stand (Tripathi et al. 2004).

The contribution of an individual species to the trees scape and the potential provision of ecosystem services can be indicated through a biomass index.

$$B = \frac{\text{Tree count} \times \text{Final height score}}{100}$$

where, species with small (5–9 m), medium (9–18 m) and large (>18 m) height are denoted with scores of 1, 2 and 3 respectively (Jim and Chen 2009; Bhat et al. 2016).

16.3 Results and Discussion

The analysis of all 25 sample plots showed that 1080 trees of 82 species belonging to 33 families were present. Out of the 25 sample plots nine, six, and ten plots were classified as rural, transitional, and urban in which 49, 39 and 56 species were observed, respectively. Thus, among the three domains, the highest number of species was found on the urban plots. The high tree species diversity within the urban domain was also confirmed by the diversity indices.

Shannon–Wiener index ranged between 0.63 and 2.48 with an average of 1.92. However, it also revealed that the urban plots had higher diversity followed by rural and transitional plots (Table 16.2). Similar results were observed in the study by Nagendra and Gopal (2010) on Shannon–Wiener index for trees of Bengaluru with maximum value of 2.68. Simpson's index ranged between 0.27 and 0.89 with an average of 0.77. The significant difference in diversity across the rural–urban gradient was observed due to higher species richness and abundance in the urban plots when compared to the rural and transitional plots (Table 16.2). The plot U11S and U16S of transitional domain had maximum diversity in terms of both the indices. Our results were in contrast to the study by Vakhlamova et al. (2014) along an urban to rural

Table 16.2 Species diversity along southern rural–urban interface of Bengaluru, India

Plot No.	Plot ID	No. of trees	No. of species	*H'*	*Cd*	*D*	*R*	*e*
1	U1S	22	10	2.031	0.169	0.831	2.912	0.882
2	U2S	61	15	2.151	0.170	0.830	3.406	0.794
3	U3S	52	15	2.156	0.172	0.828	3.543	0.796
4	U4S	19	10	2.059	0.163	0.837	3.057	0.894
5	U5S	16	7	1.450	0.352	0.648	2.164	0.745
6	U6S	38	14	2.076	0.213	0.787	3.574	0.787
7	R7S	164	18	2.143	0.171	0.829	3.530	0.742
8	U7S	45	10	1.287	0.486	0.514	2.364	0.559
9	U8S	34	12	2.129	0.171	0.829	3.119	0.857
10	U9S	46	15	2.169	0.167	0.833	3.657	0.801
11	U10S	53	17	2.381	0.138	0.862	4.030	0.840
12	R11S	13	5	1.311	0.325	0.675	1.559	0.815
13	U11S	41	17	2.481	0.118	0.882	4.578	0.876
14	U12S	11	5	1.390	0.289	0.711	1.668	0.863
15	U13S	31	11	1.939	0.220	0.780	2.912	0.809
16	U14S	52	18	2.381	0.155	0.845	4.302	0.824
17	U16S	73	18	2.462	0.108	0.892	3.962	0.852
18	R16S	67	12	1.683	0.301	0.699	2.616	0.677
19	U17S	28	11	2.093	0.163	0.837	3.001	0.873
20	R17S	29	7	1.635	0.239	0.761	1.782	0.840
21	U18S	91	21	2.299	0.168	0.832	4.434	0.755
22	U19S	38	15	2.362	0.133	0.867	3.849	0.872
23	R19S	27	5	0.625	0.731	0.269	1.214	0.388
24	R20S	6	5	1.561	0.222	0.778	2.232	0.970
25	U20S	23	8	1.747	0.217	0.783	2.233	0.840

gradient which showed increasing plant diversity towards the periphery. These results were also similar to those of the study by Aronson et al. (2014) which showed higher species richness towards the older city with both exotic and native tree species.

The Pielou's evenness index varied from 0.39 to 0.97 with an average value of 0.79. The tree species were more or less evenly distributed along the plots but the value decreased for the plots U7S of the urban domain, and R19S of the rural domain. Nevertheless, the trend of species distribution pattern slightly decreased towards the transitional and the rural plots. Margalef's index of richness varied from plot to plot but showed a decreasing trend towards transitional and rural sampling plots. They varied from 1.21 to 4.58 with an average index value of 3.03. This reflected the presence of higher numbers of diverse species in urban plots. Few rural and transitional plots (U18S and U11S in particular) had higher species richness when compared to other urban plots (Table 16.2).

The biomass index varied from 0.050 to 0.659 with the total biomass index being 11.992 (Table 16.4, see annex). The proportional biomass index (%) showed that exotic species contributed more towards total biomass along the transect. This discloses the adaptive characteristics of exotic tree species to the local environment. Studies by Mutlu et al. (2017), Roy and Mukherjee (2011), Vakhlamova et al. (2016), Nagendra and Gopal (2010, 2011) also showed similar results of dominance of exotic species over native species in urban regions. Similar tree species such as *Samanea saman, Peltophorum pterocarpum, Pongamia pinnata, Michelia champaca, Delonix regia, Lagerstroemia speciosa, Eucalyptus* sp. (Nagendra and Gopal 2010), *Bauhinia* sp., *and Tabebuia* sp. (Nagendra and Gopal 2011) were found along the streets of Bengaluru. On the other hand, the study conducted by Bhat et al. (2016) showed that native species contributed most to the total biomass.

A total of 21 tree species were common in urban, transitional and rural domains with 11 native and 10 exotic species. According to the IVI values, tree species like *Mangifera indica, Artocarpus heterophyllus, Pongamia pinnata, Tamarindus indica, Cocos nucifera, Azadirachta indica, Tectona grandis, Melia dubia, Grevillea robusta, Muntingia calabura, Psidium guajava, Moringa oleifera* and *Leucaena leucocephala* were found with higher values along the southern transect (Table 16.3).

Cocos nucifera was commonly found in all the three zones with high frequencies due to the favourable climatic conditions and its native consumption and religious significance. It was also observed that trees such as *Pongamia pinnata, Azadirachta indica, Tecoma stans*, and *Muntingia calabura* were planted predominantly in urban areas. In rural areas, multi-purpose and timber yielding tree species such as *Melia dubia, Grevillea robusta, Gliricidia sepium*, and *Tectona grandis* were preferred. Species like *Pongamia pinnata, Leucaena leucocephala, Tamarindus indica*, and *Melia dubia* were profusely planted in transition areas showing the features of both urban and rural domain.

Table 16.3 Importance Value Index of all domains along the studied transect in Bengaluru, India

Species	Urban	Transitional	Rural	Uses
Alangium salviifolium		5.06		Timber
Albizia lebbeck		6.99		Ornamental
Artocarpus heterophyllus	14.66	6.28	5.66	Fruit
Azadirachta indica	6.83		22.53	Bio-pesticide
Carica papaya		9.03		Fruit
Citrus medica		7.50		Fruit
Cocos nucifera	57.34	73.15	59.95	Fruit/nuts
Commiphora caudata			6.50	Medicinal/Ornamental
Ficus drupacea	7.92			Religious
Ficus religiosa	7.47			Religious
Ficus tinctoria	5.02	5.42		Religious
Gliricidia sepium			5.50	Fodder
Grevillea robusta		5.27	11.81	Poles
Leucaena leucocephala		11.63		Ornamental
Mangifera indica	12.58	10.89	10.31	Fruit
Melia dubia		15.14	15.66	Timber
Moringa oleifera	18.21	28.58	16.65	Edible
Muntingia calabura	12.84	5.79		Fruit
Pongamia pinnata	19.76	18.69	7.91	Seed oil
Psidium guajava	5.17		6.51	Fruit
Pterygota alata	12.15			Ornamental
Samanea saman			5.09	Ornamental
Tamarindus indica		9.46	17.26	Fruit
Tecoma stans	7.52			Ornamental
Tectona grandis	16.95	6.71	24.67	Timber

16.4 Conclusions

This study presented consolidated data on tree species identified along the urban–rural gradient in the Southern part of Bengaluru. The IVI values represented the preference for multi-purpose, fruit yielding tree species over exotic tree species along the entire transect. It also showed the presence of more diverse species in urban as compared to transitional and rural plots. In addition to it, the exotic tree species contributed more towards the biomass index. The study shows that fast growing and flowering exotic tree species are preferred over native species under urban conditions in order to make the premises more attractive and appealing. However, native tree species also need to be planted in combination with exotics to contribute to combating air pollution, enhancing carbon sequestration, nurturing other species, and maintaining urban ecology. Accordingly, more tree species should be planted in transitional and rural areas to cope with the moving anthropogenic pressures and stress from the urban development.

Acknowledgements The authors are thankful to the funding agency, Department of Biotechnology (DBT), New Delhi for funding the project study. Authors are also grateful to the Director, Institute of Wood Science and Technology (IWST) for providing all the necessary facilities for the study. The authors express sincere thanks to Prof. Christoph Kleinn, project head of FOR2432 subproject C02, for critically reviewing the manuscript and providing valuable suggestions for substantial improvement in the manuscript.

Annex

See Annex Table 16.4.

Table 16.4 List of species with the biomass index along the southern gradient of Bengaluru

S. No.	Species	Origin	Family	Total No. of trees	Density	Frequency	IVI	Biomas Index	Biomass Index (%)
1	*Acacia arabica (L.) Delile*	Native	Fabaceae	1	0.04	4	0.437	0.063	0.525
2	*Acacia catechu*	Native	Fabaceae	2	0.08	4	0.584	0.079	0.655
3	*Acacia ferruginea*	Native	Fabaceae	2	0.08	8	1.433	0.304	2.535
4	*Acacia leucophloea (Roxb.) Willd*	Native	Fabaceae	1	0.04	4	0.627	0.210	1.751
5	*Acacia senegal (L.) Willd*	Native	Fabaceae	1	0.04	4	0.459	0.226	1.885
6	*Aegle marmelos*	Native	Rutaceae	1	0.04	4	0.511	0.206	1.718
7	*Alangium salviifolium*	Native	Cornaceae	5	0.20	16	2.265	0.218	1.818
8	*Alastonia macrophylla*	Exotic	Apocynaceae	1	0.04	4	0.538	0.052	0.434
9	*Albizia amara (Roxb.) Boiv*	Native	Fabaceae	9	0.36	12	2.533	0.207	1.725
10	*Albizia lebbeck (L.) Benth*	Native	Fabaceae	5	0.20	20	3.416	0.304	2.532
11	*Albizia procera (Roxb.)Benth*	Native	Fabaceae	1	0.04	4	1.044	0.068	0.567
12	*Annona reticulata L.*	Exotic	Annonaceae	5	0.20	16	1.972	0.065	0.542
13	*Annona squamosa L.*	Exotic	Annonaceae	2	0.08	8	0.899	0.068	0.563

(continued)

Table 16.4 (continued)

S. No.	Species	Origin	Family	Total No. of trees	Density	Frequency	IVI	Biomas Index	Biomass Index (%)
14	*Araucaria cunninghamii*	Exotic	Araucariaceae	1	0.04	4	0.742	0.576	4.803
15	*Areca catechu*	Exotic	Arecaceae	11	0.44	16	2.618	0.182	1.518
16	*Artocarpus heterophyllus*	Native	Moraceae	26	1.04	48	9.031	0.086	0.719
17	*Azadirachta indica*	Native	Meliaceae	54	2.16	48	12.860	0.087	0.729
18	*Bauhinia purpurea*	Native	Fabaceae	1	0.04	4	0.531	0.052	0.434
19	*Butea monosperma (Lam.) Taub*	Native	Fabaceae	1	0.04	4	0.592	0.190	1.584
20	*Carica papaya*	Exotic	Caricaceae	20	0.80	28	4.674		
21	*Cascabela thevetia (L.) Lippold*	Exotic	Apocynaceae	2	0.08	4	0.578		
22	*Cassia siamea Lam*	Native	Fabaceae	6	0.24	8	1.425	0.090	0.749
23	*Citrus maxima*	Exotic	Rutaceae	3	0.12	8	1.026	0.069	0.575
24	*Citrus medica*	Native	Rutaceae	5	0.20	16	1.871		
25	*Cocos nucifera L.*	Exotic	Arecaceae	269	10.76	84	61.793	0.268	2.232
26	*Commiphora caudata*	Native	Burseraceae	10	0.40	24	4.025	0.074	0.614
27	*Cordia dichotoma*	Native	Boraginaceae	1	0.04	4	0.502	0.078	0.650
28	*Cordia monoica Roxb*	Native	Boraginaceae	1	0.04	4	0.456	0.080	0.667
29	*Couroupita guianensis*	Exotic	Lecythidaceae	1	0.04	4	0.730	0.190	1.584
30	*Dalbergia latifolia Roxb*	Native	Fabaceae	1	0.04	4	0.532	0.256	2.135
31	*Dead*			2	0.08	8	0.926	0.079	0.659
32	*Delonix elata (L.) Gamble*	Native	Fabaceae	4	0.16	8	1.416		
33	*Dilonix regia*	Exotic	Fabaceae	2	0.08	8	0.986	0.075	0.625
34	*Eucalyptus hybrid*	Exotic	Myrtaceae	2	0.08	8	0.972	0.545	4.541
35	*Euphorbia synadenium*	Native	Euphorbiaceae	14	0.56	12	2.629	0.051	0.421
36	*Euphorbia tirucalli*	Exotic	Euphorbiaceae	1	0.04	4	0.472	0.067	0.559
37	*Ficus benghalensis*	Native	Moraceae	2	0.08	4	2.082	0.287	2.393
38	*Ficus drupacea*	Native	Moraceae	3	0.12	4	2.724	0.271	2.262
39	*Ficus elastica*	Native	Moraceae	1	0.04	4	0.609	0.079	0.659

(continued)

Table 16.4 (continued)

S. No.	Species	Origin	Family	Total No. of trees	Density	Frequency	IVI	Biomas Index	Biomass Index (%)
40	*Ficus glomerata*	Native	Moraceae	4	0.16	16	2.456	0.081	0.673
41	*Ficus religiosa*	Native	Moraceae	2	0.08	4	2.570	0.659	5.491
42	*Ficus tinctoria*	Native	Moraceae	4	0.16	16	3.484	0.221	1.843
43	*Gliricidia sepium*	Exotic	Fabaceae	17	0.68	8	2.708	0.072	0.604
44	*Grevillea robusta*	Exotic	Proteaceae	24	0.96	20	6.911	0.333	2.779
45	*Lagerstroemia speciosa (L.) Pers*	Native	Lythraceae	2	0.08	4	0.611	0.054	0.450
46	*Lawsonia inermis L*	Native	Lythraceae	1	0.04	4	0.438	0.058	0.484
47	*Leucaena leucocephala*	Exotic	Fabaceae	27	1.08	16	5.356	0.203	1.696
48	*Mangifera indica*	Native	Anacardiaceae	36	1.44	68	11.340	0.057	0.475
49	*Manilkara zapota*	Exotic	Sapotaceae	9	0.36	16	2.389	0.067	0.555
50	*Mathuca longifolia (J.Konig) J.F.Macbr*	Native	Sapotaceae	3	0.12	8	1.355	0.057	0.475
51	*Melia dubia*	Native	Meliaceae	29	1.16	32	9.747	0.305	2.542
52	*Michelia champaca*	Native	Magnoliaceae	4	0.16	12	1.596	0.084	0.700
53	*Miliusa tomentosa (Roxb.) J.Sinclair*	Native	Annonaceae	1	0.04	4	0.470	0.088	0.734
54	*Morinda coreia Buch.-Ham*	Native	Rubiaceae	3	0.12	12	1.555	0.060	0.500
55	*Moringa oleifera*	Native	Moringaceae	96	3.84	72	19.832	0.065	0.538
56	*Morus alba*	Exotic	Moraceae	4	0.16	16	2.046	0.061	0.511
57	*Muntingia calabura*	Exotic	Muntingiaceae	27	1.08	44	7.767	0.061	0.505
58	*Murraya koenigii*	Native	Rutaceae	10	0.40	28	3.475	0.050	0.418
59	*Nyctanthes arbor-tristis L.*	Native	Oleaceae	2	0.08	8	0.891	0.055	0.454
60	*Peltophorum pterocarpum*	Native	Fabaceae	7	0.28	4	1.321	0.234	1.949
61	*Persea americana*	Exotic	Lauraceae	1	0.04	4	0.583	0.186	1.551
62	*Phyllanthus acidus*	Exotic	Phyllanthaceae	7	0.28	20	2.526	0.064	0.530
63	*Phyllanthus emblica*	Native	Phyllanthaceae	1	0.04	4	0.453	0.086	0.717

(continued)

Table 16.4 (continued)

S. No.	Species	Origin	Family	Total No. of trees	Density	Frequency	IVI	Biomas Index	Biomass Index (%)
64	*Pithocellbium dulce*	Exotic	Fabaceae	2	0.08	8	1.023	0.231	1.926
65	*Podocarpus nerifolius D. Don*	Native	Podocarpaceae	1	0.04	4	0.442	0.071	0.592
66	*Polyalthia longifolia*	Exotic	Annonaceae	9	0.36	12	2.290	0.082	0.686
67	*Pongamia pinnata*	Native	Fabaceae	61	2.44	52	14.555	0.078	0.649
68	*Psidium guajava*	Exotic	Myrtaceae	19	0.76	40	5.491	0.051	0.428
69	*Pterygota alata (Roxb.) R. Br*	Native	Malvaceae	20	0.80	4	4.241	0.302	2.517
70	*Ricinus communis*	Exotic	Euphorbiaceae	3	0.12	8	0.999	0.052	0.436
71	*Samanea saman*	Exotic	Fabaceae	5	0.20	16	3.388	0.218	1.821
72	*Santalum album L.*	Native	Santalaceae	4	0.16	16	1.771	0.064	0.536
73	*Swietenia mahagoni*	Exotic	Meliaceae	5	0.20	16	2.353	0.217	1.811
74	*Syzygium cumini*	Native	Myrtaceae	4	0.16	12	1.611	0.087	0.728
75	*Tabebuia impetiginosa (Mart. ex DC.) Standl*	Exotic	Bignoniaceae	1	0.04	4	0.441		
76	*Tabebuia rosea*	Exotic	Bignoniaceae	3	0.12	8	1.123	0.214	1.785
77	*Tamarindus indica*	Exotic	Fabaceae	16	0.64	20	9.798	0.246	2.049
78	*Tecoma stans*	Exotic	Bignoniaceae	13	0.52	12	2.779	0.201	1.679
79	*Tectona grandis*	Native	Lamiaceae	99	3.96	32	18.102	0.210	1.749
80	*Terminalia catappa*	Native	Combretaceae	7	0.28	12	2.095	0.068	0.568
81	*Terminalia chebula Retz*	Native	Combretaceae	1	0.04	4	0.450	0.332	2.769
82	*Thespesia populnea*	Native	Malvaceae	3	0.12	4	0.662	0.079	0.655
83	*Unknown*			2	0.08	4	0.549	0.061	0.509
84	*Vitex negundo L.*	Native	Lamiaceae	1	0.04	4	0.437	0.068	0.567
				1080	**43.20**	**1212**	**300**	**11.992**	

Note For few species biomass index couldn't be calculated as the average tree height was less than 5 m

References

Aronson MF, La Sorte FA, Nilon CH et al (2014) A global analysis of the impacts of urbanization on bird and plant diversity reveals key anthropogenic drivers. Proc R Soc b: Biol Sci 281(1780):1–8

Basavaraja KR (1984) History and culture of Karnataka: early times to unification. Chalukya Publications, Delhi

Bhat AA, Sharma BK, Jain AK (2016) Diversity and composition of roadside tree species at a metropolitan city of India. Imp J Interdiscip Res 2(8):66–73

Census of India (2011) Provisional population totals. Rural Urban distribution. Paper 2, vol 1. http://censusindia.gov.in/2011-Common/CensusData2011.html

Colding J, Lundberg J, Folke C (2006) Incorporating green-area user groups in urban ecosystem management. Ambio 35:237–244

Jim CY, Chen WY (2009) Diversity and distribution of landscape trees in the compact Asian city of Taipei. Appl Geogr 29(4):577–587

Ludwig JA, Reynolds JF (1988) Statistical ecology. Wiley, New York

Montgomery MR (2008) The urban transformation of the developing world. Science 319(5864):761–764

Mutlu SS, Selim C, Gülçin ÜN (2017) Plant biodiversity of urban roadside trees in Antalya, Turkey. J Forestry Faculty Kastamonu Univ 17(1):80–87

Nagendra H, Gopal D (2010) Street trees in Bengaluru: density, diversity, composition and distribution. Urban Forestry Urban Green 9(2):129–137

Nagendra H, Gopal D (2011) Tree diversity, distribution, history and change in urban parks: studies in Bengaluru, India. Urban Ecosyst 14(2):211–223

Paul S, Nagendra H (2015) Vegetation change and fragmentation in the mega city of Delhi: Mapping 25 years change. Appl Geogr 58(2015):153–166

Pielou EC (1975) Ecological diversity. Wiley, New York

Roy D, Mukherjee SK (2011) Diversity of trees in Kalyani township in West Bengal. J Econ Taxon Bot 35(4):687–695

Shannon CE, Wiener W (1963) The mathematical theory of communication. University of Illinois Press, Illinois, Urbana

Simpson EH (1949) Measurement of diversity. Nature 163(4148):688–688

Stohlgren TJ (2007) Measuring plant diversity: lessons from the field. Oxford University Press, New York

Sudhira HS, Nagendra H (2013) Local assessment of Bengaluru: graying and greening in Bengaluru—impacts of urbanization on ecosystems, ecosystem services and biodiversity. In: Elmqvist T et al (eds) Urbanization, biodiversity and ecosystem services: Challenges and opportunities. Springer, Dordrecht, pp 75–91

Tripathi KP, Tripathi S, Selven T et al (2004) Community structure and species diversity of Saddle Peak forests in Andaman Island. Trop Ecol 45(2):241–250

Vakhlamova T, Rusterholz HP, Kanibolotskaya Y et al (2014) Changes in plant diversity along an urban-rural gradient in an expanding city in Kazakhstan, Western Siberia. Landscape Urban Plan 132(2014):111–120

Vakhlamova T, Rusterholz HP, Kamkin V (2016) Recreational use of urban and suburban forests affects plant diversity in a Western Siberian city. Urban Forestry Urban Green 17(2016):92–103

Chapter 17
Biodiversity of Birds Along the Rural–Urban Interface of the Indian Megacity Bengaluru

Ingo Grass, Arne Wenzel, Varsha Raj, Suryanarayana Subramanya, Catrin Westphal, and Teja Tscharntke

Abstract Urban areas are increasing in number and extent worldwide. Few other anthropogenic land uses alter landscapes in a more persistent fashion; however, the effects of urbansiation on biodiversity remain poorly studied. Here, we studied bird communities along the rural–urban interface of the Indian megacity Bengaluru. Birds were assessed with point counts on 36 vegetable-growing farms situated within three urbanisation stages (urban, peri-urban and rural) and across three seasons (monsoon, winter, summer). We recorded 4617 individuals comprising 149 species. Bird richness was highest in peri-urban farms and lowest in rural areas, while urban areas were intermediate. Bird abundance did not differ among urbanisation stages. Increasing amount of impervious area in the surrounding landscape reduced bird richness and

I. Grass (✉) · A. Wenzel · T. Tscharntke
Agroecology, Georg-August-University Göttingen, 37077 Göttingen, Germany
e-mail: ingo.grass@uni-hohenheim.de

A. Wenzel
e-mail: arne.wenzel@agr.uni-goettingen.de

T. Tscharntke
e-mail: ttschar@gwdg.de

I. Grass
Ecology of Tropical Agricultural Systems, University of Hohenheim, 70599 Stuttgart, Germany

V. Raj
Wildlife Institute of India, Dehradun 560065, India
e-mail: varsha0404@gmail.com

S. Subramanya
Post Harvest Engineering & Technology, University of Agricultural Sciences Bangalore, Bengaluru 560065, India
e-mail: subbu.subramanya@gmail.com

C. Westphal
Functional Agrobiodiversity, Georg-August-University Göttingen, 37077 Göttingen, Germany
e-mail: catrin.westphal@agr.uni-goettingen.de

T. Tscharntke
Center of Biodiversity and Sustainable Land Use, Georg-August-University Göttingen, 37077 Göttingen, Germany

E. Hoffmann et al. (eds.), *The Rural-Urban Interface*, The Urban Book Series,
https://doi.org/10.1007/978-3-030-79972-4_17

abundance, with the greatest effect up to 20% impervious area, levelling off beyond this threshold. Bird community composition on urban farms was distinct from rural areas, while peri-urban areas supported species from both species pools. In conclusion, urbanisation and particularly impervious areas threaten Bengaluru's bird diversity. Peri-urban areas that mark the transition from rural to urban areas are hotspots of bird diversity. City planning should aim to retain green areas and avoid densification, particularly in the peri-urban interface where urbanisation happens most rapidly.

Keywords Biodiversity assessment · Land cover · Bird species diversity · Bird abundance · Bird community patterns · Agroecology

17.1 Introduction

Today, 55% of the world's population live in urban areas and it is expected that by 2050 population growth and shift of residence will have added another 2.5 billion people to urban areas (United Nations 2018). In an increasingly urbanised world, maintaining agriculture within and around cities is receiving a growing interest. Agriculture proximal to urban areas is often characterised by higher land prices necessitating more intensively managed farming systems, such as by increasing agrochemical input and resource use as well as by changes in crop patterns and production systems (Chaps. 3, 7, and 11). This, in turn, often reduces the associated and planned agrobiodiversity (Sudhira and Nagendra 2013; Chap. 19). Furthermore, high amounts of built-up area and fragmentation of remaining green spaces (Chap. 2) reduce landscape-wide species pools and habitat connectivity (Faeth et al. 2011).

Birds are an ecologically important and well-studied taxonomic group that is prevalent across all major land uses worldwide, including cities. Although in general urbanisation can be expected to impoverish bird communities (Batáry et al. 2018), the effects of urbanisation are highly variable and context-dependent. Numerous studies report reduced bird richness in urban as compared to non-urban or rural areas (Clergeau et al. 2006; Sandström et al. 2006). However, given intermediate levels of urbanisation (e.g. built-up areas intermixed with high amount of green spaces), urban areas can maintain bird richness as high or even higher than rural areas (Batáry et al. 2018). For bird abundance, the dominance of a few species adapted to urban areas can result in higher numbers inside cities than in rural areas (Chace and Walsh 2006; Minor and Urban 2010; Faeth et al. 2011; Njorge et al. 2014).

Despite an increasing number of studies comparing bird communities in urban versus rural areas, there is a scarcity of studies along gradients of urbanisation (but see Chamberlain et al. 2017). This is problematic, as bird species responses to urbanisation may in fact be nonlinear, with more species being present in the transition from peri-urban to urban areas, as compared to the rural to peri-urban interface (Batáry et al. 2018). In addition, most studies on bird communities in urban environments are still conducted in temperate regions, although both the rate of urbanisation and its negative effects on biodiversity are likely to be even stronger in the highly biodiverse subtropics and tropics.

Here, we focus on changes in bird diversity with urbanisation in the metropolitan area of the Indian megacity Bengaluru. We find pronounced changes in bird communities, with the highest bird species richness in transitional (peri-urban) areas but similarly high abundance in urban, peri-urban and rural areas. The landscape context strongly affected bird communities, whereby high amounts of impervious areas in the surroundings strongly and nonlinearly reduced local bird richness as well as abundance on farms.

17.2 Materials and Methods

17.2.1 Study Design

Fieldwork was conducted in 2017 in Bengaluru, India. The research design comprised two transects starting close to the city's centre and ending in the rural surroundings, with a northern and a southwestern direction, respectively. Within each transect, we selected 18 vegetable-growing farms (36 farms in total) (Fig. 17.1). Farms were selected according to the urbanisation stage of the surrounding urban context: six farms were located within an urban context, six farms in transitional areas and six farms within a rural context, following the classification of Hoffmann et al. (2017). As an additional measure of urbanisation, we used satellite data (Landsat, 10 m × 10 m resolution) to infer the proportion of impervious area (built-up, sealed surfaces) in a 500 m radius around each selected farm.

17.2.2 Bird Surveys

We used point counts to assess the bird community of each farm. Our surveys from September 2017 to May 2018 covered the three seasons that characterise farming in Bengaluru: monsoon (September, October), winter (December, January) and summer (April, May). Each farm was sampled twice during each season (36 farms × 3 seasons × 2 point counts = 216 point counts in total). A point count consisted of 15 min acoustic and visual surveys of the bird community within a 50 m radius from the centre of each selected farm. All point counts were done in the morning hours (06:00–08:00) and done by the same observer (Varsha Raj).

17.2.3 Statistical Analysis

We aggregated all data from the point counts at the site level ($n = 36$), with species richness defined as the total number of observed species across all point counts per

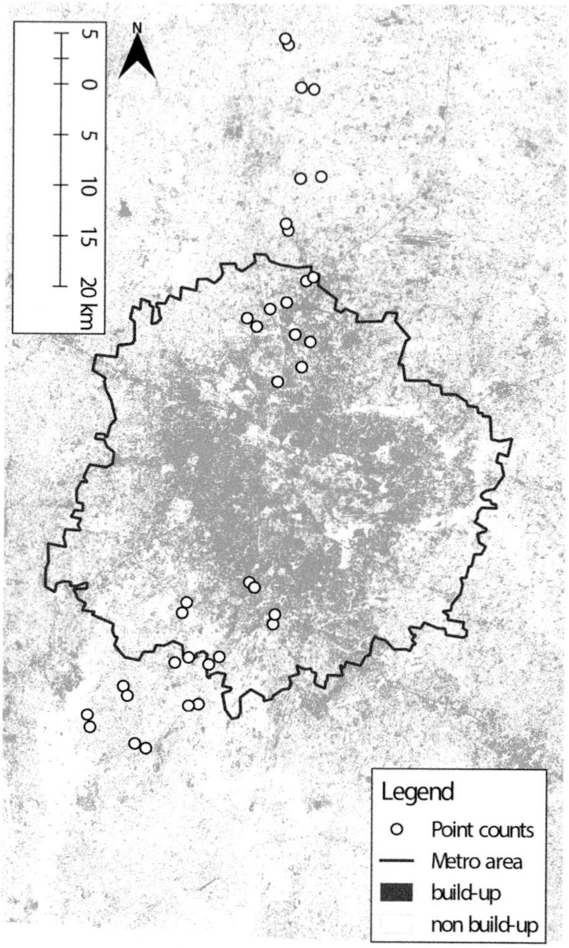

Fig. 17.1 Map of the Bengaluru region, India, and localities of field sites. Studies of bird diversity were conducted on 36 vegetable farms. Urban areas with high built-up amount are shown in grey, non-built-up areas (e.g. green spaces) are shown in white colour

site. For species abundance, we used the maximum number of detected individuals per species and site (not the total sum of all detected individuals across all surveys) to avoid inflated abundances due to repeated detections. We used generalised linear models with negative binomial error distributions to model variation in species richness and species abundance of birds. Thereby, we used multiple regressions with urbanisation stage (urban, peri-urban, rural) and the amount of impervious area in 500 m radius surrounding the 36 farms (percentage value) as predictors. Tukey's test was used to infer significant differences ($p < 0.05$) in predicted means of bird richness and abundance among urbanisation stages. In addition, we used non-metric dimensional scaling (NMDS) to investigate variation in community composition of farms situated in urban, peri-urban and rural areas.

17.3 Results

In total, we recorded 4617 bird individuals comprising 149 species during our surveys. Of these, 125 bird species predominantly used terrestrial and 22 species used aquatic habitats.

The three most abundant species were the common mynah (*Acridotheres tristis*, 150 individuals), the rose-ringed parakeet (*Psittacula krameri*, 132 individuals) and the large-billed crow (*Corvus macrorhynchos*, 130 individuals). All detected bird species are native to South India, while 24 species are winter visiting migrants. Only three aquatic species, the oriental darter, the painted stork and the black-headed ibis (*Anhinga melanogaster*, *Mycteria leucocephala* and *Threskiornis melanocephalus*) are listed as 'near threatened' in the IUCN Red List (IUCN 2019).

The degree of urbanisation strongly affected bird diversity. The predicted bird species richness was highest in farms classified as situated in peri-urban areas and lowest in rural areas, while urban areas were intermediate (Fig. 17.2a). In contrast, bird abundance did not differ among urbanisation stages (Fig. 17.2c). The landscape context surrounding the farms strongly affected local bird diversity: both species

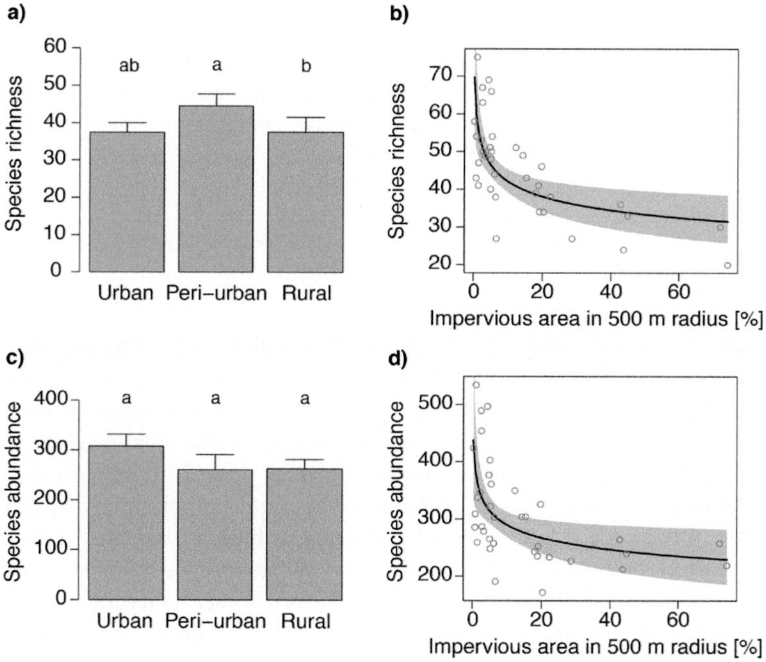

Fig. 17.2 Changes in bird diversity of vegetable-growing farms with urbanisation in Bengaluru, India. Bird species richness (**a**) but not abundance (**c**) differs among urbanisation stages (mean + 1 SE; letters indicate significant differences). Both richness (**c**) and abundance (**d**) decrease nonlinearly with increasing amount of impervious area surrounding the vegetable farms (mean predictions and 95% confidence intervals)

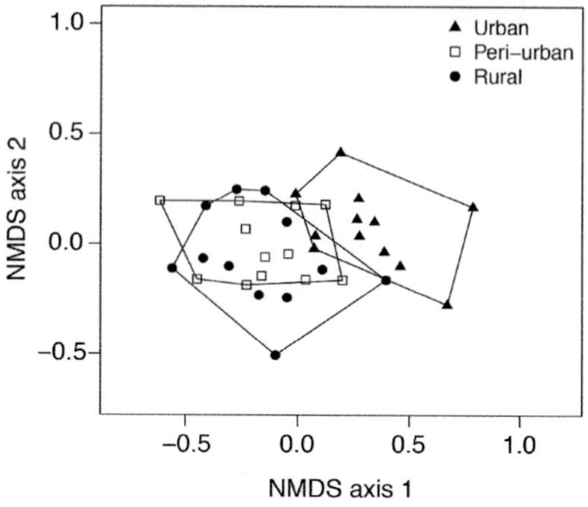

Fig. 17.3 Bird community composition on 36 farms (symbols) situated in urban, peri-urban and rural urbanisation stages. Closely plotted symbols indicate similar species compositions of farms, while distantly plotted symbols indicate differences in composition

richness and abundance of birds strongly decreased with higher amounts of impervious area in a 500 m radius surrounding the study sites (Fig. 17.2b, d), with the greatest effect up to 20% impervious area, levelling off beyond this threshold.

The composition of bird communities strongly differed among farms (stress value of NMDS: 0.19) and was largely explained by the urbanisation stage ($r^2 = 0.38$, $p < 0.001$). Bird community composition on farms located in urban areas was distinct from those in rural areas. In contrast, while generally more similar to rural communities, some peri-urban communities ranged in between, indicating mixture of species from urban and rural species pools (Fig. 17.3).

17.4 Discussion

We found pronounced changes in bird communities with urbanisation in the metropolitan area of Bengaluru. Vegetable-growing farms situated in peri-urban areas supported the highest bird richness. In contrast, bird richness was reduced in rural areas and strongly declined with the proportion of impervious areas in a 500 m radius surrounding the farms, with greatest losses occurring up to 20% impervious area.

Urbanisation commonly leads to impoverished bird communities (Batáry et al. 2018); the effects of urbanisation, however, are highly variable and context-dependent. High amounts of green spaces and urban sprawl, in contrast to densification, can maintain bird richness as high as or even higher than rural areas (Batáry et al. 2018). Correspondingly, we found the highest bird richness on vegetable farms situated in peri-urban areas that are characterised by higher amounts of green areas than the dense urban core of Bengaluru. Furthermore, the higher bird richness in peri-urban as compared to rural areas may result from a mixture of rural and urban species pools in this transitional zone. Urban and rural bird communities can be expected to be functionally distinct, for example with respect to habitat associations or migratory strategy (Minor and Urban 2010). Hence, follow-up analysis of their functional attributes may provide deeper insights into their potential mixing in peri-urban areas. All urbanisation stages supported similarly high bird abundance, contrasting with studies emphasising the dominance of few adapted species in urban areas (Faeth et al. 2011; Njorge et al. 2014). Likewise, the meta-analysis by Batáry et al. (2018) indicates that species abundance of birds peaks in peri-urban (termed 'suburban' by Batáry et al. 2018) areas; our study, however, supported this only for patterns in species richness but not abundance.

In addition to the urbanisation stage, the surrounding landscape context strongly affected local bird diversity. We found a steep, nonlinear loss in bird richness and abundance with increasing amount of impervious area, which was particularly pronounced and steep from 0 to 20% impervious area in the surrounding landscape. These findings indicate that even small losses of green areas lead to strongly impoverished bird communities. Specialist birds (forest specialists, insectivores) are particularly sensitive to habitat loss and fragmentation and thus avoid urban matrices (Kennedy et al. 2010). Also in our study, specialists like the jungle myna (*Acridotheres fuscus*) or the western cattle egret (*Bubulcus ibis*) were only observed on vegetable farms with low impervious area (<20%). Conservation and restoration of green areas as well as their connectivity are thus key to maintaining diverse bird communities in Bengaluru.

17.5 Conclusions

With 149 species recorded on 36 investigated vegetable-growing farms, the Indian megacity Bengaluru harbours astonishingly high bird diversity. Peri-urban areas that mark the transition from rural to urban areas in particular are hotspots of bird diversity. However, increasing urbanisation and impervious areas that result from the sealing of surfaces and built-up densification are strong threats to Bengaluru's bird diversity. City planning, thus, should aim at retaining green areas and avoid urban densification, particularly in the peri-urban interface where urbanisation happens most rapidly.

Acknowledgements IG and TT were supported by the Deutsche Forschungsgemeinschaft (DFG) through grant numbers GR 4844/1-1 (IG) and TS 45/34-1 (FOR2432). CW is grateful for funding by the DFG (project number 405945293). We thank Prof. Dr. Vasuki Belavadi and the staff of the Entomology Department at the University of Agriculture Sciences Bangalore for guidance and logistic support.

References

Batáry P, Kurucz K, Suarez-Rubio M, Chamberlain DE (2018) Non-linearities in bird responses across urbanization gradients: a meta-analysis. Glob Change Biol 24(3):1046–1054
Chace JF, Walsh JJ (2006) Urban effects on native avifauna: a review. Landscape Urban Plan 74(1):46–69
Chamberlain D, Kibuule M, Skeen R, Pomeroy D (2017) Trends in bird species richness, abundance and biomass along a tropical urbanization gradient. Urban Ecosyst 20(3):629–638
Clergeau P, Croci S, Jokimäki J, Kaisanlahti-Jokimäki ML, Dinetti M (2006) Avifauna homogenisation by urbanisation: analysis at different European latitudes. Biol Conserv 127(3):336–344
Faeth SH, Bang C, Saari S (2011) Urban biodiversity: patterns and mechanisms. Ann NY Acad Sci 1223(1):69–81
Hoffmann EM, Jose M, Nölke N et al (2017) Construction and use of a simple index of urbanisation in the rural-urban interface of Bangalore, India. Sustainability 9:2146
IUCN. The IUCN Red List of Threatened Species (2019) Report No. 2019-3. http://www.iucnredlist.org
Kennedy C, Marra PP, Fagan WF et al (2010) Landscape matrix and species traits mediate responses of Neotropical resident birds to forest fragmentation in Jamaica. Ecol Monogr 80(4):651–669
Minor E, Urban D (2010) Forest bird communities along a gradient of urban development. Urban Ecosyst 13(1):55–71
Njorge JB, NdaNg'ang'a PK, Natuhara Y (2014) The pattern of distribution and diversity of avifauna over an urbanizing tropical landscape. Urban Ecosyst 17(1):61–75
Sandström UG, Angelstam P, Mikusinski G (2006) Ecological diversity of birds in relation to the structure of urban greenspace. Landscape Urban Plan 77(1–2):39–53
Sudhira HS, Nagendra H (2013) Local assessment of Bengaluru: graying and greening in Bengaluru—impacts of urbanization on ecosystems, ecosystem services and biodiversity. In: Elmqvist T et al (eds) Urbanization, biodiversity and ecosystem services: challenges and opportunities. Springer open access. http://link.springer.com/book/10.1007%2F978-94-007-7088-1
United Nations (2018) Revision of world urbanization prospects. www.un.org/development/desa/publications/2018-revision-of-world-urbanization-prospects.html

Chapter 18
Bee Diversity in the Rural–Urban Interface of Bengaluru and Scope for Pollinator-Integrated Urban Agriculture

Chethana V. Casiker, B. Jagadishakumara, G. M. Sunil, K. Chaithra, and M. Soubadra Devy

Abstract Cities all over the world are expanding rapidly and eating into their rural peripheries. Pollination is a key ecosystem service essential for the majority of crops. As natural vegetation and agricultural land are lost to urban expansion, it becomes imperative to preserve pollinators within cityscapes. We examined the bee community in Bengaluru to ascertain whether the city can support pollination services in the light of the growing interest in urban agriculture. Using pan traps, we sampled bees in 18 sites spread across the rural–urban interface of North Bengaluru. Background floral composition was surveyed within a buffer area around each site. Urban sites recorded higher species richness of both bees and flowering plants, than the adjoining transition and rural sites. The correlation observed between bees and flowering plants is in line with other studies that highlight the role of the floral community in shaping bee community dynamics. Further, it is likely that urban regions are favourable to bees as they lack large swaths of chemically intensive monocultures which are common in rural areas. The study indicates that the city is capable of supporting initiatives such as urban farming and locavory that could help improve green infrastructure and liveability of urban landscapes.

Keywords Biodiversity assessment · Bee species diversity · Floral resources · Land cover · Pollination · Bee community patterns

---

C. V. Casiker · B. Jagadishakumara · G. M. Sunil · K. Chaithra · M. S. Devy (✉)
Ashoka Trust for Research in Ecology and the Environment, Bengaluru 560064, India
e-mail: soubadra@atree.org

C. V. Casiker
e-mail: chethanacasiker@atree.org

B. Jagadishakumara
e-mail: jagdish.kumar@atree.org

G. M. Sunil
e-mail: sunil.gm@atree.org

K. Chaithra
e-mail: chaithra.k@atree.org

E. Hoffmann et al. (eds.), *The Rural-Urban Interface*, The Urban Book Series,
https://doi.org/10.1007/978-3-030-79972-4_18

18.1 Introduction

By 2050, about 68% of the world's population is expected to be living in urban areas (UNDESA 2018) and by 2030, India's city-dwelling population will double to 590 million (McKinsey and Company 2010). There is increasing need to understand how urban growth and the demands that follow can be managed in sustainable ways through nature-based solutions. This is critical to ensure that cities remain liveable in the future.

Recent studies indicate that the assumptions that we may have regarding the nature of urban areas need to be reconsidered. For instance, Shastri et al. (2017) analysed the changes in surface temperature of 84 Indian cities using satellite data. They found that urban regions in some parts of India were actually cooler than nearby non-urban areas. This was attributed to higher evaporation in cities resulting from high human consumption of water and increased evapotranspiration from planted trees and grasses. There is growing evidence that cities with better green infrastructure and more water bodies are less impacted by the urban heat island effect (Gill et al. 2007; Saaroni et al. 2018). Green infrastructure includes spaces in and around urban areas that deliver multiple functions, as they provide ecological, economic and social benefits (Beyer et al. 2014), promote sustainable living and support appropriate urban development (Mell 2009). So far, very few cities have integrated urban agriculture into development planning although this would further increase green infrastructure, and thereby improve the microclimate of the area (Qiu et al. 2013).

Often discussed in the context of liveable cities is the concept of urban renewal with emphasis on 'locavory', that is, eating locally grown food and thereby reducing the carbon footprint. The practice of locavory contributes to multiple dimensions of green infrastructure as it enhances food production in the cities with the support of biodiversity, in particular pollinators. A report from the United Nations Food and Agriculture Organization revealed that an estimated 15 to 20% of global food production already takes place in or near urban areas (Armar-Klemesu 2000). Worldwide, approximately 800 million people earn their livelihood as urban farmers (Orsini et al. 2013). Clearly, urban agriculture is a fast-growing area of research. Nevertheless, the ability of urban ecosystems and urban dwelling pollinators to provide pollination services remains understudied.

In this chapter, we examine green infrastructure of the South Indian city of Bengaluru in the context of locavory. The city, in recent times, has seen the growth of terrace/balcony gardening and increasing interest in the cultivation of vegetables and fruits, plants that are largely dependent on pollinators (Matteson and Langellotto 2009). Using bees as indicators, we examine plant assemblages across various land uses to determine the kind of pollinator support that they offer. Based on the findings, we propose the nature of pollinator supported agriculture that might be suitable for urban agriculture in Bengaluru. We address the following questions: Is there a compositional change in bee fauna between the city and the surrounding transition zones? Does the flora within the city influence bee composition? Will the existing bee flora support the emerging urban agriculture in the city?

18.2 Materials and Methods

18.2.1 Site Selection

The study area was the rural–urban interface in the north of the city of Bengaluru. Along a gradient of urbanisation, based on the rural–urban index developed by Hoffmann et al. (2017), 18 sites were chosen and categorised as urban, transitional and rural (6 sites each). The centre of every site was an agricultural farm. Buffers with radii of 250 and 500 m were laid around each site. These distances were determined based on typical bee flight distances (Gathmann and Tscharntke 2002).

18.2.2 Pan Trap Sampling

We sampled for bees using pan traps (Westphal et al. 2008), an effective method, well-suited for large scale, passive sampling of flying insects. We chose pan traps of yellow and blue colours as they captured the highest number of specimens (Bates et al. 2011), while white was discarded after initial field trials.

Pan trap sampling was carried out between April and June 2018, as this period comprised the local active crop growing season. The weather during the sampling period was mostly sunny, with conditions favourable for bee sampling before the onset of the monsoon season.

Bees were collected from the farm sites, as well as the region within the buffers. About 90 traps were laid at each site, amidst different types of vegetation that were representative of that area. Under these conditions, species counts had reached saturation (Fig. 18.1). Plastic cups filled with soapy water were left in field for 48 h (Bates et al. 2011), after which the specimens were collected and preserved in ethanol

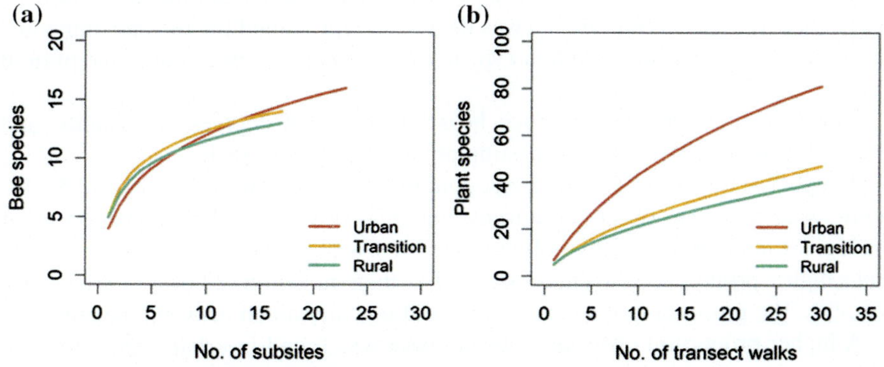

Fig. 18.1 Saturation curves for sampling of bee species (**a**) and floral resources (**b**) across the three areas representing different urbanisation stages in Bengaluru, India

for sorting and identification. All insect samples were sorted to order level. Bee specimens were identified to morphospecies level. Sample specimens were cleaned, pinned and preserved for reference.

18.2.3 Assessment of Background Flora

In order to study the response of the bee community to the background floral composition, we conducted floral surveys in the study sites. As bees are sensitive to floral resources in the landscape (Hines and Hendrix 2005), the floral assessment was carried out within the 500 m radius buffer at every site. In order to get an accurate assessment of the background flora at the time of sampling, the floral assessment was conducted at the same time as the setting of pan traps (Fig. 18.1). Five locations were chosen at random at each site, and data on flora was collected by means of a 100 m transect walk. As a measure of floral abundance, the number of open flowers was recorded across multiple species of flora by a method of approximation (Cusser et al. 2015). The floral species, colour of the flower and stage of flowering was also recorded. Flowering plants were identified up to the species level.

18.3 Results

18.3.1 Bees Species Composition Across the Rural–Urban Gradient

We collected a total of 1727 individual bees from 19 species (Fig. 18.2). The most abundant genus was *Lasioglossum* (62.7% of all bees collected), followed by *Apis florea* (19.2%; Fig. 18.2). *Lasioglossum* clearly emerged as the dominant genus in the study area. The genus *Homalictus* was the rarest, represented by just one individual recorded in an urban site. *Seladonia* sp. and *Tetragonula* sp. were not found in rural sites.

Based on literature we classified bees into different degrees of sociality, and grouped them according to their nesting preferences (Table 18.1).

A higher abundance of social bees, comprising 45.8% eusocial bees and 52.4% semi-social bees, was observed in transition areas (Fig. 18.3). Urban sites reported the highest number of socially polymorphic bees, namely, *Lasioglossum*. Rural sites had a higher proportion of solitary bees (49%) than transition (22.6%) or urban sites (28.3%). The proportion of subsocial bees did not vary much between the areas.

A higher proportion of ground-nesting bees was found in urban areas (42.3%), while rural areas had the lowest proportion (24.7%). On the contrary, urban sites had the lowest share of bees which nested in trees/buildings (20.8%), compared to transition (45.8%) and rural sites (33.4%; Fig. 18.4).

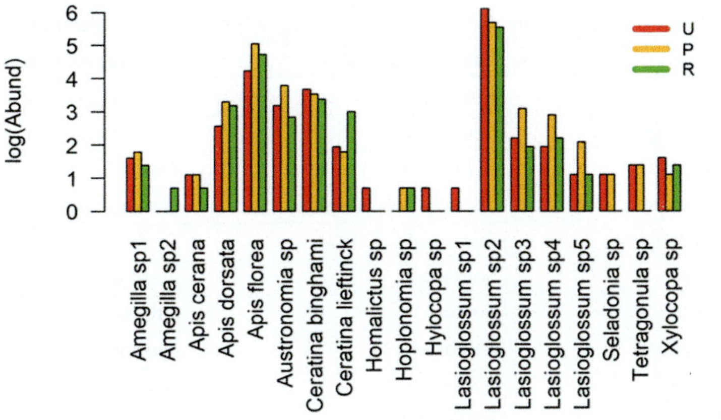

Fig. 18.2 Distribution of bee species abundance across the areas representing different urbanisation stages in Bengaluru, India (U = urban, P = transitional, R = rural)

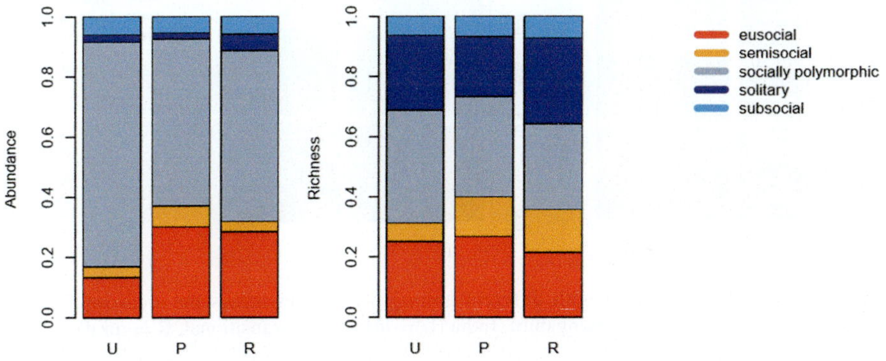

Fig. 18.3 Distribution of bees with different degrees of sociality across the areas representing different urbanisation stages in Bengaluru, India (U = urban, P = transitional, R = rural)

18.3.2 Composition of Floral Abundance of Bee-Supporting Flora Across the Rural–Urban Gradient

The plant family *Fabaceae* was observed to have the highest representation among bee-supporting flora in the study area (Fig. 18.5). This was followed by *Verbenaceae, Myrtaceae* and *Asteraceae* families. Flowering plants from the families *Amaryliidaceae, Bignoniaceae, Muntingiaceae, Papavaraceae, Poaceae* and *Scrophulariaceae* were found only in urban sites. Transition areas recorded high floral abundance from families *Verbenaceae, Myrtaceae, Fabaceae* and *Rosaceae. Convolvulaceae* was recorded only in rural sites. *Chrysanthemum* sp., *Jacaranda mimosifolia, Peltophorum pterocarpum* and *Sida acuta* were recorded only in urban sites. Species such as *Tamarindus indica* and *Rosa* sp. were recorded in both urban

Table 18.1 Degree of sociality and nesting preferences of bees (based on literature)

Genus/species	Degree of sociality	Nesting preference
Apis sp.	Eusocial	Trees, buildings
Austronomia sp.	Semisocial	Ground/soil
Hoplonomia sp.	Semisocial	Ground/soil
Lasioglossum sp.	Socially polymorphic	Ground/soil
Ceratina binghami	Subsocial	Dead wood
Xylocopa sp.	Solitary	Dead wood
Amegilla sp.	Solitary	Ground/soil
Ceratina lieftinck	Solitary	Dead wood
Homalictus sp.	Solitary	Ground/soil
Seladonia sp.	Wide range	Ground/soil
Megachile sp.	Solitary	Dead wood

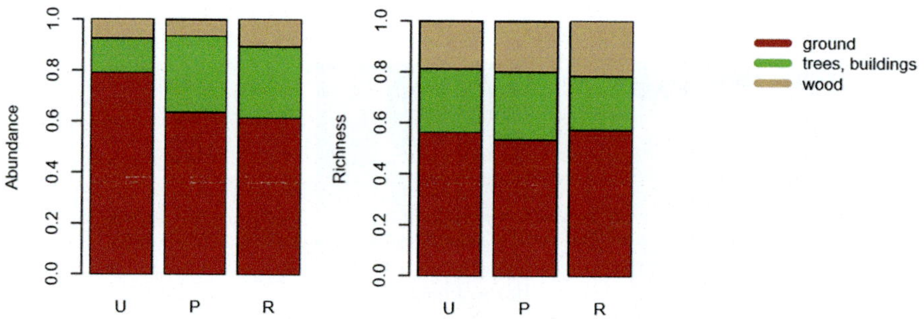

Fig. 18.4 Distribution of bees with different nesting preferences across the areas representing different urbanisation stages in Bengaluru, India (U = urban, P = transitional, R = rural)

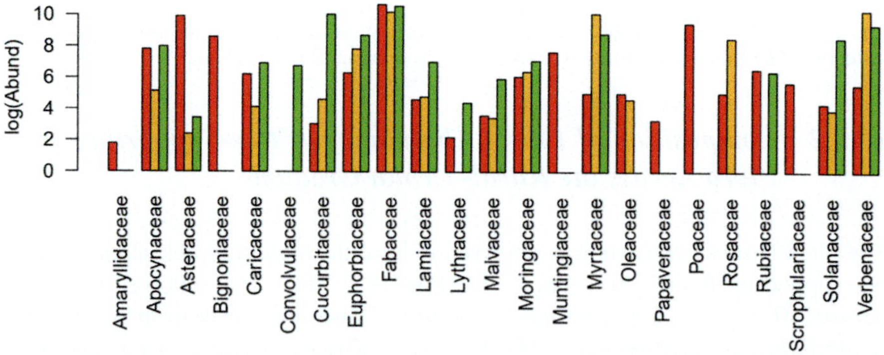

Fig. 18.5 Distribution of floral abundance of families of bee-supporting plants across the areas representing different urbanisation stages in Bengaluru, India (red denotes urban sites, yellow denotes transition sites, green denotes rural sites)

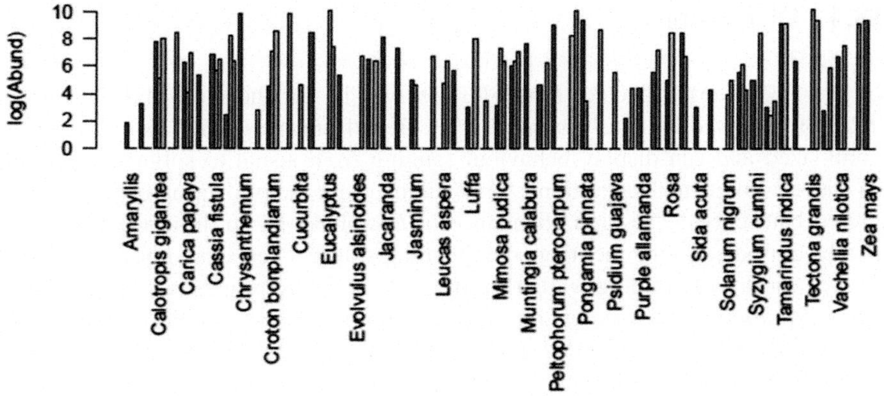

Fig. 18.6 Distribution of floral abundance of bee-supporting plants across the three areas representing different urbanisation stages in Bengaluru, India (red denotes urban sites, yellow denotes transition sites, green denotes rural sites)

and transition sites. *Calotropis gigantea, Carica papaya, Cassia fistula, Croton bonplandianum* and *Mimosa pudica* were found across the gradient (Fig. 18.6).

18.3.3 Interaction of Bees and Surrounding Flora

Species richness of flowering plants was highest in urban sites by a wide margin, thus indicating that urban sites support higher diversity of plants, which was confirmed by the Shannon diversity index (Table 18.2). Bees showed a correspondingly high species richness in urban sites (Table 18.2), but the Shannon diversity index for bees was highest in transition areas.

Table 18.2 Comparison of flowering plant species richness and bee species richness across the three areas representing different urbanisation stages in Bengaluru, India

Category	Urban	Transitional	Rural
Species richness			
Flowering plants	81	47	40
Bees	17	15	14
Shannon diversity index			
Flowering plants	2.86	2.31	2.67
Bees	1.16	1.61	1.48

18.4 Discussion

Of the 1727 bees collected, *Lasioglossum* was by far the most abundant genus (62.7%), followed by *Apis florea* (19.2%). Bees from the genus *Lasioglossum* are small-sized and can display behaviour ranging from social to solitary depending on the environmental conditions. This might prove helpful in adjusting to new, altered landscapes. The high abundance of *Lasioglossum* sp. in urban sites could also be attributed to the fact that some members of this genus are known to nest in the soft mortar of building walls (Sirohi et al. 2015). Rare genera such as *Homalictus, Seladonia* and *Tetragonula* were encountered only in urban and transition sites. Social bees were recorded in highest numbers in transition areas. Urban sites reported the highest number of socially polymorphic bees (*Lasioglossum*). Rural sites had the highest proportion of solitary bees. Urban areas reported a high proportion of ground-nesting bees, while rural areas had the lowest. Disturbances caused by tilling and other agricultural activities in rural areas might be a reason for low numbers of below-ground nesting bees (Williams et al. 2010). Urban sites had the lowest numbers of bees which nested in trees or buildings compared to transition and rural sites.

In line with the view that loss of native habitat and human land use cause a decline in pollinators, many studies have shown a declining trend in bee composition with urbanisation (Winfree et al. 2009; Bates et al. 2011; Choate et al. 2018). However, several studies also indicate the opposite view that urban habitats are favourable for pollinators, particularly bees. Sirohi et al. (2015) reported a highly urbanised town to be more diverse and abundant in solitary and primitively eusocial bees than meadows and nature reserves. A study by Winfree et al. (2007) found that bee abundance and species richness were negatively and not positively, associated with the extent of forest cover in the surrounding landscape, suggesting that moderate disturbance of natural habitat may in fact be conducive for conservation of certain species of bees. Our study, which shows that urban sites have high bee species richness, is in agreement with this observation. We also recorded higher bee abundances in urban (634 individuals) and transition (616 individuals) sites compared to rural sites (477 individuals).

It is important to note that rural areas do not necessarily have large areas of natural vegetation. Despite the low overall built up area, pollinators and pollination service are negatively impacted as the rural landscape loses complexity and becomes increasingly dominated by monocultures (Landis 2017). Agricultural fields are intensively farmed and possibly have high levels of pesticide usage and disturbance to native biota. These factors are expected to have a negative effect on bees (Goulson et al. 2015). Cities, on the other hand, have small pockets of greenery with a variety of plants, including a greater number of exotic species. Despite having lower green cover, urban areas boast of vegetation that is highly diverse and can provide habitat for forage and shelter to different species of bees. The diversity of flora in urban spaces and the ensuing differences in flowering patterns lead to year-round availability of flowers in contrast with crop fields in rural settings. This is likely to render urban landscapes more attractive to bees, as flowering is not restricted to any particular season. This is consistent with the drastic difference in floral richness between urban and rural areas in our study. Our results showed that the species richness of flowering plants in city areas was nearly double that of the rural peripheries. Urban areas had the highest Shannon-diversity of flowering plants as well.

Our finding that bee community dynamics are strongly correlated with floral community should come as no surprise as floral diversity and nectar resource diversity are known to influence bee diversity (Potts et al. 2003). Abundance, richness, and spatial distribution of floral resources have been acknowledged as strong factors in shaping bee community structure in urban settings (Plascencia and Philpott 2017). Several analyses and meta-analyses of urban insect pollinators have shown that floral forage consistently correlates with pollinator health (Hall et al. 2017; Potts et al. 2003). Our findings support this as both flowering plants and bees had highest species richness in urban sites. Flowers of trees such as the Blue Jacaranda *(Jacaranda mimosifolia*), copperpod (*Peltophorum pterocarpum*) and wireweed (*Sida acuta*), which are bee-pollinated, were recorded in high numbers in urban sites. Flowering plants such as lilies and *Chrysanthemum* were recorded only in urban sites. Tamarind (*Tamarindus indica*) and rose (*Rosa* spec.) were recorded in both urban and transition sites. Giant Milkweed (*Calotropis gigantean*), Papaya (*Carica papaya*), Indian Laburnum (*Cassia fistula*), Jungle Tulsi (*Croton bonplandianum*) and Touch-me-not (*Mimosa pudica*) were found across the gradient.

Agricultural production relies to a large extent on animal-mediated pollination. Globally, insect pollinators influence the production of 39 of the 57 major crops by affecting the quality and quantity of fruit and seed (Klein et al. 2007). In developing and transition countries such as India, agricultural production has been shown to be 50% more pollinator dependent than that of developed countries (Aizen et al. 2009). This, coupled with the global trend of increase in land under cultivation of pollinator-dependent crops (Aizen et al. 2008), provides impetus for integrating pollinators and pollination service in agriculture.

As urban agriculture gains importance, especially in fast-urbanising regions of the world, the ability of pollinators to cope with urbanisation and provide pollination service merits attention (Wenzel et al. 2020). This leads us to the question of whether existing bee flora in the cities can support gardening and locavory efforts in the future. Ghazoul (2006) pointed out that diverse floral displays are expected to attract a greater variety of pollinators. As several studies have indicated that urban areas are home to diverse plants (McKinney 2008), it is likely that they are capable of supporting a good population of pollinators. In return, a healthy diversity of bees supports pollination of multiple plant species (Lowenstein et al. 2015). Hall et al. (2017) opine that pollinators put high priority and high-impact urban conservation within reach, owing to their relatively small functional requirements with respect to habitat range, life cycle, and nesting. This highlights the need and potential for conservation of pollinators in cities. With increasing awareness of the benefits of consuming fresh, pesticide-free and locally-grown food, cities all over the world are experiencing a surge of interest in activities such as kitchen gardening and terrace gardening. Bengaluru is no exception. Indeed, many food plants can be grown relatively easily within the climatic conditions and space constraints faced in the city (Aiholli and Bargavi 2018; Jaganmohan et al. 2012; Vazhacharickal 2014).

18.5 Conclusions

Our results highlight the potential that urban landscapes have for supporting bees and pollination service. Managing cities for safeguarding biodiversity along with improving green infrastructure by integrating locavory with city planning can help citizens move towards more sustainable urbanisation. Our study is based on samples collected in a single season. A natural extension of our study would be to conduct continuous, year-round sampling across multiple seasons and years to get a better understanding of the bee community in the landscape.

Acknowledgements We thank Priyadarsanan Dharma Rajan and Seena N. Karimbumkara for their help and guidance in insect identification, and Ronita Mukherjee for help during the project design phase. We thank Seema Purushothaman for her valuable inputs. We further thank Arne Wenzel for the selection of study sites and support extended. This work was supported by the Department of Biotechnology (DBT), Government of India, as part of the Indo-German collaboration agreement between DBT, India and the German Science Foundation (DFG).

References

Aiholli SM, Bargavi MTB (2018) Perception and acceptance of rooftop farming by residents in Bangalore, India

Aizen MA, Garibaldi LA, Cunningham SA, Klein AM (2008) Long-term global trends in crop yield and production reveal no current pollination shortage but increasing pollinator dependency. Curr Biol 18(20):1572–1575

Aizen MA, Garibaldi LA, Cunningham SA, Klein AM (2009) How much does agriculture depend on pollinators? Lessons from long-term trends in crop production. Ann Bot 103(9):1579–1588

Armar-Klemesu M (2000) Urban agriculture and food security, nutrition and health. Growing cities, growing food. Urban agriculture on the policy agenda, pp 99–118

Baldock KC, Goddard MA, Hicks DM et al (2019) A systems approach reveals urban pollinator hotspots and conservation opportunities. Nat Ecol Evol 3(3):363

Bates AJ, Sadler JP, Fairbrass AJ et al (2011) Changing bee and hoverfly pollinator assemblages along an urban-rural gradient. PloS one 6(8)

Beyer KM, Kaltenbach A, Szabo A et al (2014) Exposure to neighbourhood green space and mental health: evidence from the survey of the health of Wisconsin. Int J Environ Res Public Health 11(3):3453–3472

Choate BA, Hickman PL, Moretti EA (2018) Wild bee species abundance and richness across an urban–rural gradient. J Insect Conserv 22(3–4):391–403

Cusser S, Neff JL, Jha S (2015) Land use change and pollinator extinction debt in exurban landscapes. Insect Conserv Divers 8(6):562–572

Gathmann A, Tscharntke T (2002) Foraging ranges of solitary bees. J Anim Ecol 71(5):757–764

Ghazoul J (2006) Floral diversity and the facilitation of pollination. J Ecol 94(2):295–304

Gill SE, Handley JF, Ennos AR et al (2007) Adapting cities for climate change: the role of the green infrastructure. Built Environ 33(1):115–133

Goulson D, Nicholls E, Botías C, Rotheray EL (2015) Bee declines driven by combined stress from parasites, pesticides, and lack of flowers. Science 347(6229):1–10

Hall DM, Camilo GR, Tonietto RK (2017) The city as a refuge for insect pollinators. Conserv Biol 31(1):24–29

Hines HM, Hendrix SD (2005) Bumble bee (Hymenoptera: Apidae) diversity and abundance in tallgrass prairie patches: effects of local and landscape floral resources. Environ Entomol 34(6):1477–1484

Hoffmann EM, Jose M, Nölke N et al (2017) Construction and use of a simple index of urbanisation in the rural–urban interface of Bangalore, India. Sustainability 9(11):2146

Jaganmohan M, Vailshery LS, Gopal D et al (2012) Plant diversity and distribution in urban domestic gardens and apartments in Bangalore, India. Urban Ecosyst 15(4):911–925

Klein AM, Vaissiere BE, Cane JH, Steffan-Dewenter I, Cunningham SA, Kremen C, Tscharntke T (2007) Importance of pollinators in changing landscapes for world crops. Proc R Soc b: Biol Sci 274(1608):303–313

Landis DA (2017) Designing agricultural landscapes for biodiversity-based ecosystem services. Basic Appl Ecol 18:1–12

Lowenstein DM, Matteson KC, Minor ES (2015) Diversity of wild bees supports pollination services in an urbanized landscape. Oecologia 179(3):811–821

Matteson KC, Langellotto GA (2009) Bumble bee abundance in New York City community gardens: implications for urban agriculture. Cities Environ (CATE) 2(1):5

McKinney ML (2008) Effects of urbanization on species richness: a review of plants and animals. Urban Ecosyst 11(2):161–176

McKinsey Global Institute (2010) India's urban awakening: building inclusive cities, sustaining economic growth

Mell IC (2009) Can green infrastructure promote urban sustainability? In: Proceedings of the institution of civil engineers-engineering sustainability, vol 162(1), pp 23–34. Thomas Telford Ltd

Orsini F, Kahane R, Nono-Womdim R et al (2013) Urban agriculture in the developing world: a review. Agron Sustain Dev 33(4):695–720

Plascencia M, Philpott SM (2017) Floral abundance, richness, and spatial distribution drive urban garden bee communities. Bull Entomol Res 107(5):658–667

Potts SG, Vulliamy B, Dafni A et al (2003) Linking bees and flowers: how do floral communities structure pollinator communities? Ecology 84(10):2628–2642

Qiu GY, Li HY, Zhang QT et al (2013) Effects of evapotranspiration on mitigation of urban temperature by vegetation and urban agriculture. J Integr Agric 12(8):1307–1315

Saaroni H, Amorim JH, Hiemstra JA (2018) Urban Green Infrastructure as a tool for urban heat mitigation: survey of research methodologies and findings across different climatic regions. Urban Clim 24:94–110

Shastri H, Barik B, Ghosh S (2017) Flip flop of day-night and summer-winter surface urban heat island intensity in India. Sci Rep 7(1):1–8

Sirohi MH, Jackson J, Edwards M (2015) Diversity and abundance of solitary and primitively eusocial bees in an urban centre: a case study from Northampton (England). J Insect Conserv 19(3):487–500

UNDESA— United Nations Department of Economic and Social Affairs (2018) 2018 revision of world urbanization prospects

Vazhacharickal PJ (2014) Balcony and terrace gardens in Urban greening and local food production: scenarios from Mumbai Metropolitan Region (MMR), India. Inter J Food Agric Veterinary Sci 4(2):149–162

Wenzel A, Grass I, Belavadi VV, Tscharntke T (2020) How urbanization is driving pollinator diversity and pollination—a systematic review. Biol Conserv 241:108321

Westphal C, Bommarco R, Carré G (2008) Measuring bee diversity in different European habitats and biogeographical regions. Ecol Monogr 78(4):653–671

Williams NM, Crone EE, T'ai HR (2010) Ecological and life-history traits predict bee species responses to environmental disturbances. Biol Cons 143(10):2280–2291

Winfree R, Griswold T, Kremen C (2007) Effect of human disturbance on bee communities in a forested ecosystem. Conserv Biol 21(1):213–223

Winfree R, Aguilar R, Vázquez DP et al (2009) A meta-analysis of bees' responses to anthropogenic disturbance. Ecology 90(8):2068–2076

Chapter 19
Dynamics of Field Margin Vegetation with Changing Agricultural Landscape Across a Rural–Urban Interface

Sunil Nautiyal, Mrinalini Goswami, Ellen Hoffmann, and Puneeth Shivakumar

Abstract Field margins are areas of natural or semi-natural vegetation located adjacent to agricultural fields. They form the interface between different crops and to other land use and land cover types. In most analyses of cropping patterns field margin vegetation (FMV) is neither accounted for nor is it a part of so-called natural ecosystems. Nevertheless, FMV provides important ecosystem services, is integrated in the farming practices of traditional smallholder agriculture, and contributes to farmers' income. The area, composition and functions of FMV change as agroecosystems are transformed under progressing urbanisation, as encountered in the periphery of Bengaluru, India. This study is a first attempt to systematically assess FMV in such a context. It focuses on different characteristics of FMV along a spatial rural–urban gradient, and its interrelation with changes in crop and farm management.

Keywords Landscape structure · Field margins · Vegetation assessment · Farmer survey · Biodiversity · Ecosystem services · Agroecology

19.1 Introduction

Agroecosystems are managed ecosystems from which humans derive economic benefits in the form of food, fodder, fiber or other produce and services. As such

S. Nautiyal (✉) · M. Goswami · P. Shivakumar
Centre for Ecological Economics and Natural Resources (CEENR),
Institute for Social and Economic Change (ISEC) Nagarabhavi, Bengaluru 560072, India

M. Goswami
e-mail: goswamimrinalini@gmail.com

P. Shivakumar
e-mail: s.puneeth42@gmail.com

E. Hoffmann
Organic Plant Production and Agroecosystems Research in the Tropics and Subtropics, University of Kassel, 37213 Witzenhausen, Germany
e-mail: ellen.hoffmann@uni-kassel.de

E. Hoffmann et al. (eds.), *The Rural-Urban Interface*, The Urban Book Series,
https://doi.org/10.1007/978-3-030-79972-4_19

they are a constituent of both, human society and natural ecosystems. Agroecosystems may comprise various components such as croplands, grazing land, livestock, neighboring forest and wetlands, as well as the interfaces between them, and yet they can be considered a single functional unit. The biodiversity of agricultural landscapes comprises species and genetic diversity from both agricultural and non-agricultural land and water habitats. Ecosystems with higher diversity exhibit higher resistance and resilience, enabling them to withstand undesirable changes or to recover following disturbance (Altieri and Nicholls 2004). Compared to natural ecosystems, agroecosystems have higher fluidity and vulnerability, more spatiotemporal differences, and reduced biodiversity (Zhu et al. 2012). Changes in agroecosystems are closely interrelated with the socioeconomic development of the community depending on their provisioning and regulating services.

Field margins form an integral part of the non-cropped habitats of agroecosystems, and enrich their biodiversity by harboring a variety of floral and faunal species (Marshall 1987; Rands and Whitney 2011). The ecologically most significant part of non-crop biodiversity in agricultural landscapes was found in field boundaries with hedgerows of modified wood lots (Garratt et al. 2017), which contain tree, shrub and herbaceous vegetation layers. Such field boundaries form a barrier to wind erosion and non-point pollution, play a role in keeping the field water balance, and control nutrient losses. They provide habitats to pests but also support to natural pest controlling agents. Despite their small size as compared to the cropped area, they show a great richness of organisms and provide multiple services. Farm boundaries receive seeds at greater densities than open farms and pastures, which makes them species-rich and worthwhile maintaining for their numerous ecological and socioeconomic benefits (Pandey 2002). Smith et al. (2008) summarised the ecological functions of field margins in their *biodiversity value* (they increase the species density in an agroecosystem), *conservation value* (they provide habitats for rare or endangered species) and *functional value* (they enhance ecosystem services such as pest control and decomposition). Many studies have demonstrated increased abundance of natural enemies and more effective biological control where crops are bordered by wild vegetation (Altieri 1999). The presence of a weedy filed margin may be more effective in enhancing biological control and pest suppression than annual flowering field margins (Balzan and Moonen 2014). Continuous removal of field margin vegetation negatively impacts regulatory landscape functions and the efficiency of agriculture. Beside the many biotic and abiotic factors encountered at the landscape scale, cropping practices and farm types have a profound effect on the quantity, type, and diversity of field margin vegetation, which in turn has a feedback impact on the crops.

In the ancient days, agricultural lands were delineated by field margins to represent the land ownerships and protect the fields from wild animals (Marshall 2004). The benefits of FMV were traditionally known to local farmers, and they therefore retained, maintained and used the vegetation on the field boundary. In the course of intensification and commercialization of farming, however, field margin vegetation was increasingly perceived as having low economic value while occupying potentially productive space. It was not included in the farm management practices (DeSnoo 1994) and if not cleared, was widely neglected. India is still lagging behind in the management strategies of field margins and extensive research has to be carried out to achieve biodiversity conservation on field margins in an economically and logistically efficient manner.

Structure and composition of field margin vegetation are determined by landscape parameters and land use intensity (Billeter et al. 2008). The latter may include cropping pattern, nutrient inputs, amount of livestock, farm fragmentation, intensity of herbicide and pesticide use, plot size, and distance to neighbouring fields or other land uses, area of semi-natural patches, and proximity to natural vegetation. In a study in Norway, Hovd and Skogen (2005) found that plant species composition at the field margin was affected by thickness of litter, slope, width, moisture level and type of crop in the adjacent field. Next to intensely managed agricultural fields, the diversity of flora declined (Fritch et al. 2011). Maintaining a certain width, sowing wildflowers or beneficial plant seeds, and controlling agro-chemical use are some measures that may enhance biodiversity of FMV amidst intensified farming. The strategies for FMV management have to be adapted to the agroecological features of the local context and extensive studies are thus essential for the effective management of biodiversity in agroecosystems. However, very limited research is available on the importance of FMV in Indian landscapes dominated by agriculture, especially in areas subject to growing urbansation. The broad objective of this paper therefore is to understand how the changing agricultural landscapes in a metropolitan area have affected field margin vegetation.

19.2 Materials and Methods

19.2.1 Study Area and Survey Design

This study was set up along a transect in the north of Bengaluru (N-transect) which covers ca. 142,000 ha of land (Fig. 19.1). The transect cuts across Bengaluru Urban and Bengaluru Rural districts' jurisdictions. The selection of this study area was justified by rapid urban sprawl following the establishment of IT parks and the New International Airport. The transect was divided into three zones on the basis of population density, distance from the city core and extent of subsistence agriculture and land transformation. Villages and crop fields were then selected in rural, transitional and urban zones.

The assessment combined a phyto-social inventory of the field margins with a farmer survey in order to establish the crop-FMV relationships (Fig. 19.2, Table 19.1). Along the transect, in urban, transitional, and rural zones, square sampling tiles of 1 km^2 were laid out in triplicate to document land use type and number of farms. In total 1233 farms were detected. Out of this 12% of the area in each zone were delineated for a survey and detailed assessment of FMV.

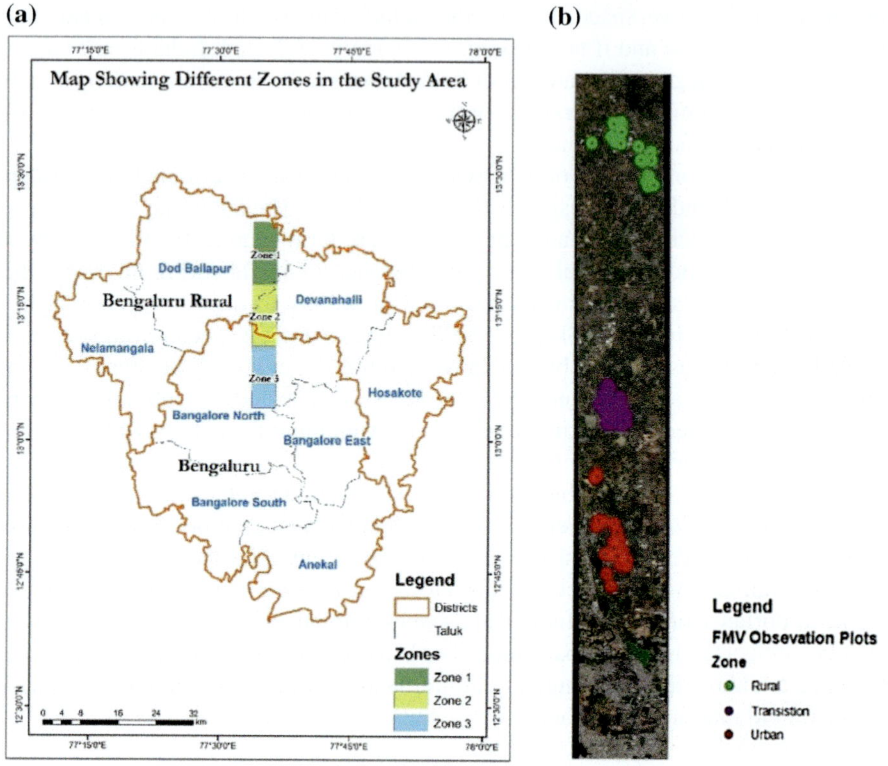

Fig. 19.1 Location of study sites in Bengaluru, India. **a** rural, urban, and transitional zones (Zone 1, 2 and 3, respectively) within the N-transect **b** Observation fields in rural (green dots), transitional (purple dots), and urban (red dots) village clusters within the zones. (Nautiyal et al. 2020)

Fig. 19.2 Exemplary delineation of FMV around two sample plots north of Bengaluru, India (Worldview 3 satellite image of 2017)

Table 19.1 Combined vegetation assessment and household survey design in the study transect of Bengaluru, India

Level of analysis	Sample (n)
Number of 1 km^2 tiles	9
Number of farms	1233
Sampling frame	12%
Household survey	150 farmers
Survey of FMV (area, trees)	131 fields (owned by the surveyed farmers)
Sub-sample for FMV assessment	56 agricultural fields

19.2.2 Household Survey on Crop and FMV Management

In total, 150 farmers owning agricultural fields selected for the vegetation assessment were included in the survey carried out in six villages. Two villages each were in the rural (Heggedahalli and Muddenahalli), transitional (Kundana and Konaghatta), and urban (Singanayakanahalli and Arekere) zone of the transect.

The questionnaire of the household survey addressed household profile, general socioeconomic information, land ownership and land use, agricultural practices and production, livestock, vegetation in the field margins, use and traditional ecological knowledge about FMV, and perceived reasons for changes in FMV. It first inquired about the present situation, and then, by recall, about past conditions in the 1990s.

19.2.3 Field Survey of Crops and Field Margin Vegetation

A total of 131 randomly selected fields owned by the surveyed farmers were identified on remote sensing images, their adjacent field margins were delineated (Fig. 19.2), and crop area and FMV area were measured. The sample comprised 49 fields in the rural village cluster (Sothenahalli, Tubagere, Melkote and Kuruvinagere), 42 fields in the village clusters of the transitional zone (Bachahalli, Pandithapura, Juttanahalli and Jalige), and 40 sample plots in the village cluster (Tharahunase, Nellukunte, Lakshinisagar and Nagadasanahalli) which was considered urban. FMV trees were identified and counted in all of them, and a subsample of 56 crop fields (21 rural, 17 transitional, 18 urban) was subject to a detailed phyto-social vegetation assessment including also shrub and herb species. For perennial tree and shrub strata all species were identified and quantified. However, for annual herbaceous vegetation only dominant species were documented during the study period. To analyse the crop-FMV interaction, crops were grouped in cereals, with finger millet (*Eleusine coracana* Gaertn.) and maize (*Zea mays* L.) as most abundant species, pulses such as horse gram (*Macrotyloma uniflorum* (Lam.) Verdec.), red gram (*Cajanus cajan* (L.) Millsp.), and lablab (*Lablab purpureus* (L.) Sweet), vegetables, plantation crops including fruit trees such as coconut (*Cocos nucifera* L.), mango (*Magnifera indica*

L.), guava (*Psidium guajava* L.), ornamentals such as flowers (*Rosa spp. Lilium spp.*) and lawn grass (*Lolium sp., Paspalum notatum, Cynodon dactylon*, or *Zoysia japonica*), and fodder such as napier grass (*Pennisetum purpureum* L.).

19.3 Results

19.3.1 Changes in the Agroecological Landscape

The farmers were well aware of profound changes in the agricultural landscape over the past three decades. They attributed it to the reduction of landholdings in the course of division of ancestral land among brothers, and to overall socioeconomic changes related to urbanisation (50% of the surveyed households). These two causes were followed by the lack of water and land sale as major concerns. Small and marginal farmers, with agricultural land of less than two hectare, preferentially stated socioeconomic change and lack of water as major reasons for change in the agricultural landscape, whereas the owners of larger farms considered market influence and division of agricultural land as major drivers of change. Almost half of the surveyed households had experienced losses of agricultural landholdings, whereas only 5% reported an increase in their agricultural land area since the 1990s. The noticeable changes in the transects in terms of agricultural activities were found to be governed by landholding size, land quality, water availability, market price of land and market-oriented crop selection.

19.3.2 Status of Field Margin Vegetation

Both the area covered by field margins and the vegetation characteristics within them are indicators of the status of FMV. Only four out of the 56 surveyed fields included in the vegetation assessment were larger than 1.3 ha, and all of those had a vegetated field margin. Of the remaining 52 fields, 21 did not have FMV. In the entire subsample, FMV area was not correlated to the size of the crop area bounded by the FMV. In fields with FMV, it accounted for about 10% of the available cultivable area.

When all mapped areas of crops and FMV were aggregated, the proportion of FMV declined from 11% in the rural zone to 5% in the transitional and 4% in urban zones, at similar total crop areas (Fig. 19.3). Cereals and plantation crops were most frequently cultivated (25 and 15 fields within the sub-sample with a total crop area of 15.9 and 10.8 ha, respectively), whereby the FMV area was higher around cereal fields (ca. 8% of the crop area) than around plantations (5%). The sharp decline of FMV area towards the transitional and urban zone was evident for both crop groups.

That the diversity and abundance of FMV was not only related to the position of the plots along the rural–urban gradient, but also depended on the crop type grown in

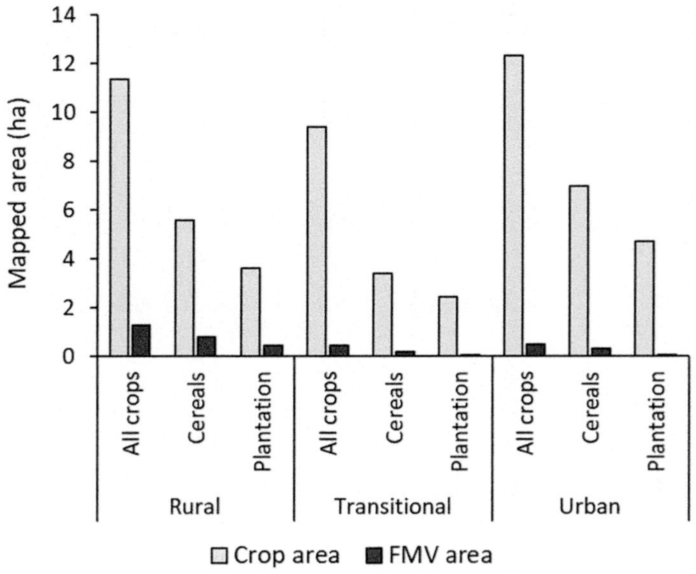

Fig. 19.3 Relations of crop and FMV area along the rural–urban gradient of Bengaluru, India

the adjacent field, was most obvious for the trees within the field margins (Table 19.2). Cereals and pulses were bordered by trees in 42%, 62% and 51% of the fields in rural, transitional and urban sections of the transect, respectively. The proportion was much lower adjacent to vegetable crops, especially in the transition zone with only 27% of plots having trees in their field margin. Grape (*Vitis vinifera* L.) cultivation stood out by having no FMV at all in the transition zone, and the lowest proportion of FMV-bordered fields in the other zones. Similar observations were made for lawn grass, which was always cultivated without maintaining any field margins.

A more comprehensive account of the crop-FMV interaction included all FMV vegetation types (trees, shrubs and herbs), as well as the area covered by the crops and their bordering FMV (Table 19.3). Broken down by rural, transitional, and urban zones, it showed the crop preferences along the urbanisation gradient, and the associated impact on area and diversity of FMV.

Among the three zones, not only the area of field margin was highest in rural areas but also the number of floral taxa found in the field margins. Area of field margin and number of plant species depict the relationship between cultivation practices and conservation of FMV.

Table 19.2 Proportion of plots having trees in the field margins for different crop groups along the rural–urban gradient and FMV tree species

		Cereals and pulses	Vegetables	Plantation		Ornamental	
Rural	Crops	Finger millet/ragi (*Eleusine coracana*), maize (*Zea mays*), redgram/pigeon pea/togari (*Cajanus cajan*)	Cucurbits (*Cucurbita sp.*), tomato (*Solanum lycopersicum*), eggplant/brinjal (*Solanum melongena*), cabbage (*Brassica oleracea*), cauliflower (*Brassica* sp.)	Arecanut (*Areca catechu*), mango (*Mangifera indica*)	Grapes (*Vitis* spp.)	Rose (*Rosa* spp.), marigold (*Tagetes* spp.), lily (*Lilium* sp.)	
	FMV%	42	60	100	42	60	
	FMV Tree species	Mango (*Mangifera indica*), *jujube* (*Ziziphus jujuba*), custard apple (*Annona squamosa*), jackfruit (*Artocarpus heterophyllus*), coconut (*Cocos nucifera*), banyan tree (*Ficus benghalensis*)	Jackfruit (*Artocarpus heterophyllus*), karanjah (*Pongamia pinnata*), coconut (*Cocos nucifera*), eucalypt/neelagiri (*Eucaluptus* sp.)	Teak (*Tectona grandis*), silver oak (*Grevillea robusta*), coconut (*Cocos nucifera*)	Coconut (*Cocos nucifera*)	Mango (*Mangifera indica*), banyan tree (*Ficus benghalensis*)	
Transitional	Crops	Jowar (*Sorghum* spp.), finger millet/ragi (*Eleusine coracana*), horsegram (*Macrotyloma uniflorum*), redgram/pigeon pea/togari (*Cajanus cajan*), mustard (*Brassica* spp.)	Cucumber (*Cucumis sativus*), coriander (*Coriandrum sativum*), chilli (*Capsicum frutescens*), tomato (*Solanum lycopersicum*), ginger (*Gingiber officinale*), beans (*Lablab* spp.)	Chikoo/sapota (*Manilkara zapota*), coconut (*Cocos nucifera*), teak (*Tectona grandis*)	Grapes (*Vitis* spp.)	Rose (*Rosa* spp.), marigold (*Tagetes* spp.)	Lawn grass

(continued)

Table 19.2 (continued)

		Cereals and pulses	Vegetables	Plantation		Ornamental	
	FMV %	62	27	50	0	67	0
	FMV Tree species	Silver oak (*Grevillea robusta*), neelagiri (*Eucalyptus*), teak (*Tectona grandis*), coconut (*Cocos nucifera*) karanjah (*Pongamia pinnata*), banyan tree (*Ficus benghalensis*), date palm (*Phoenix dactylifera*)	Coconut (*Cocos nucifera*), banyan tree (*Ficus benghalensis*)	Silver oak (*Grevillea robusta*)		Neem tree (*Azadirachta indica*), silver oak (*Grevillea robusta*)	
Urban	Crops	Finger millet/ragi (*Eleusine coracana*), maize (*Zea mays*), beans (*Lablab* spp.), horsegram (*Macrotyloma uniflorum*)	Beans (*Lablab* spp.), bitter gourd (*Momordica charantia*), cauliflower (*Brassica* sp.), tomato (*Solanum lycopersicum*)	Guava (*Psidium guajava*), mango (*Mangifera indica*), pomegranate (*Punica granatum*)	Grapes (*Vitis* spp.)	Rose (*Rosa* spp.)	Lawn grass
	FMV %	57	57	71	50	0	0
	FMV Tree species	Banyan tree (*Ficus benghalensis*), karanjah (*Pongamia pinnata*), kachnar (*Bauhinia variegata*) neem tree (*Azadirachta indica*), hebbevu (*Melia dubia*)	Teak (*Tectona grandis*), karanjah (*Pongamia pinnata*), coconut (*Cocos nucifera*)	Karanjah (*Pongamia pinnata*), coconut (*Cocos nucifera*), eucalypt/neelagiri (*Eucalyptus*), jackfruit (*Artocarpus heterophyllus*)	Teak (*Tectona grandis*), silver oak (*Grevillea robusta*), mango (*Mangifera indica*), neem tree (*Azadirchta indica*), coconut (*Cocos nucifera*)		

FMV % = percentage of plots with field margin vegetation

Table 19.3 Area under crops, FMV and associated biodiversity in the study transect of Bengaluru, India

Location			Number of species in FMV			
Crop group	Crop area total (ha)	FMV area total (ha)	Herbs	Shrubs	Trees	Total
Rural						
Cereals	5.56	0.78	20	11	8	39
Pulses	0.38	0.04	2	5	5	12
Vegetables	0.79	0[a]	5	2	0	7
Plantation	3.61	0.45	12	4	4	20
Ornamentals	1.03	0[a]	12	2	0	14
Transitional						
Cereals	3.40	0.16	14	6	10	30
Pulses	1.03	0.08	2	2	2	6
Vegetables	0.74	0.08	4	3	5	12
Plantation	2.44	0.05	1	5	5	11
Ornamentals	0.38	0.01	1	4	3	8
Fodder	1.42	0.01	3	2	3	8
Urban						
Cereals	6.97	0.29	9	7	5	21
Pulses	0.11	0[a]	6	1	0	7
Vegetables	0.43	0.08	4	3	2	9
Plantation	4.72	0.06	3	7	3	13
Ornamentals	0.08	0.04	4	3	2	9

[a]In some places the mapped FMV area was zero and there were no tree species. However, if at the very edge of those fields a few shrubs and dominant herbs species were present, they were documented as FMV

19.3.3 Use of Field Margin Vegetation

Fruit bearing and timber producing trees were found to play a crucial role in the economy of farm households, especially during crop failure and family distress situations. They support livestock, nutritional supply and faunal diversity.

The tree, shrub and herb species were categorised in ten use groups that provide different benefits, such as fruits for food and beverages, fodder, fuelwood, fibre, ornamentals, timber, and leaf litter/compost. The highest number of species (17) lend themselves to the use of plants and plant parts as food (fruits and vegetables), followed by fodder, firewood, and timber. Among the food producing trees, coconut, tamarind, papaya, wood apple, jackfruit, mango, and custard apple were mentioned throughout all three sections of the transect. The food items collected from the field margins are primarily for household consumption with only occasional selling. The percentage of households using them, however, is comparatively small (Fig. 19.4). The most

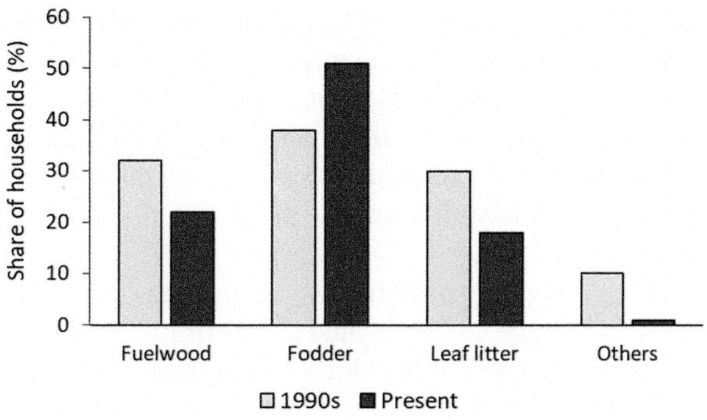

Fig. 19.4 Use of FMV-based resources and its change over time in Bengaluru, India. The use of FMV for food is included in the category 'Others'

popular uses of FMV products pertain to fuelwood, fodder and leaf litter.

The survey results indicated a decrease in households using field margins vegetation in all the use categories except for fodder (Fig. 19.4). Although there has been an increase in the number of households using fodder from 38% in 1990s to 51% at present, the quantities of resources used in all categories have decreased during that period. The increase in fodder harvesting from field boundary is another consequence of land scarcity that led to agricultural intensification and has pushed more families to adopt livestock rearing as a secondary livelihood.

19.4 Discussion

The definition of urban and transitional zones in this study differed from the SSI-based stratification (see Chap. 2). The so-called urban zone coincides with the transitional area in the other classification, and the so-called transitional villages are mostly considered rural in the other classification. However, any delineation is somewhat arbitrary, what matters is the relative positioning in the gradient. Thus, various different approaches come to the same principal results as shown repeatedly in this volume.

Agricultural field margins are places where wild species grow as well as economically and ecologically important species that are deliberately planted and conserved. Demographic and economic conditions of the changing landscape determine what the farmers want to plant, retain and use in the field margins.

The socioecological survey showed that the farmers considered Indian beech/karanjah (*Pongamia* sp.), *Carissa carandas,* star jasmine (*Jasminum multiflorum*) and farm grasses as the most abundant vegetation in the field margin, whereby farm grass is a new addition. In the 1990s *Carissa carandas, Jasminum multiflorum, Leucas aspera, Pongamia* sp. and *Lantana camara* were reported as most abundant species. Eucalypt (*Eucalyptus* sp.), coconut (*Cocos nucifera*), mango (*Mangifera indica*) and oak trees (*Grevillea robusta*), which are economically important, were mentioned as less abundant species in the field margins. The current direct usage of FMV primarily included fodder, crop protection, and fruits for household and commercial purpose. Coconut, pomegranate (*Punica granatum*), farm grass (*Pennisetum purpureum*), eucalypt, teak (*Tectona grandis*), jackfruit (*Artocarpus heterophyllus*), and mango were found to be planted in recent years. Thus, the semi-natural characteristics of field margin have shifted to plantation crops, thereby adversely affecting the multi-functionality of the agroecosystem with diverse non-crop species.

The cropping preferences along the gradient described in other chapters (Chaps. 2, 3 and 11), indicated a shift from the predominance of cereals and pulses in the rural section to horticulture and plantation towards the urban end, which was less evident here. It was shown, however, that FMV is preferentially conserved around cereal fields. As cereal cultivation decreases further, a stronger impact on the associated field margins can be expected. This not only changes the structure of the agricultural landscape, but also compromises its regulatory functions. If, on one hand, most vegetables are insect-pollinated, and, on the other hand, field margin vegetation is considered a refuge for insect populations, its removal especially around the vegetable fields is counter-productive. As for many ecological functions, this may go unnoticed until a threshold is passed and the shortage of pollinators results in noticeable yield losses.

In the present study, plantation crops, and in particular grapes cultivation has been identified as one of the most potent drivers to remove field boundary vegetation. Bengaluru Urban and Bengaluru Rural districts are among the most prominent grape growing districts of the State. As a traditional livelihood in Nandi Valley (which the study transect cuts across), grape production got a major boost from the promotion of wineries through government policy. In just seven years Karnataka has doubled the area under grape cultivation from 9700 hectares in 2007/08 to 20,400 hectares in 2013/14. Thus, the changing cultivation practices and crops are expected to jeopardize the natural and semi-natural field boundary vegetation if necessary attention is not given on time. Lawn grass, on the other hand, is a specialty crop grown in a few local hotspots, and thus the clearing of FMV due to lawn grass production has much less impact on the agricultural landscape than the widespread grape cultivation.

Finally, FMV also provides connectivity channels that link larger patches of natural vegetation. If those are increasingly interrupted, this may have indirect impact on bird populations (Chap. 17) and other associated fauna. Such interactions between biodiversity of crops, FMV, insects, and birds remain to be explored.

19.5 Conclusions

Changes in the socioeconomic status of the population driven by urbanisation sometimes lead to desirable livelihood shifts with an educated and skilled generation in a landscape that generates a positive feedback loop. The rapid changes in physical landscape conditions due to urbanisation may provide non-agricultural livelihood opportunities, but nevertheless, the population dependent on natural resources or agriculture may not be able to acquire the opportunities at the same pace, which may lead to over-exploitation of resources. This case study indicated that clearing of field margin vegetation may result in simplified and impoverished agroecosystems. External pressures for over-exploiting natural resources as well as environmental degradation have made many intensified agroecosystems loose the ecological and economic benefits of field margin vegetation.

For the well-known ethical, economic and ecological values of trees farmers don't allow to burn residues in the field if trees are present at the field margin. A comprehensive insight to these values, which are likely to change with changing agricultural landscapes, should be developed to formulate strategies for conservation of this significant component of biodiversity in agroecosystems.

The research on FMV exemplifies an unexplored dimension of agro-ecosystems that has the potential to make significant contribution in sustainable resource flow and proper conservation and management of FMV could play an essential role in mitigating the threat related to climate change. This study is a first step in attracting the attention of the stakeholders interested in investigation of socio-ecology and its sustainability.

Acknowledgements We are thankful to the Department of Biotechnology Govt. of India for financial support. This work has been carried out as part of an Indo-German research cooperation (DFG: Research Unit FOR2432/1; DBT: The Rural-Urban Interface of Bengaluru—A Space of Transitions in Agriculture, Economics, and Society.

References

Altieri MA (1999) Applying agroecology to enhance the productivity of peasant farming systems in Latin America. Environ Dev Sustain 1(3):197–217

Altieri MA, Nicholls CI (2004) An agroecological basis for designing diversified cropping systems in the tropics. J Crop Improv 11(1–2):81–103

Balzan MV, Moonen AC (2014) Field margin vegetation enhances biological control and crop damage suppression from multiple pests in organic tomato fields. Entomol Exp Appl 150(1):45–65

Billeter R, Liira J, Bailey D et al (2008) Indicators for biodiversity in agricultural landscapes: a pan-European study. J Appl Ecol 45(1):141–150

DeSnoo GR (1994) 13 Cost-benefits of unsprayed crop edges in winter wheat, sugar beet and potatoes. Unsprayed field margins: implications for environment, biodiversity and agricultural practice. Dissertation, University of Leiden

Fritch RA, Sheridan H, Finn JA et al (2011) Methods of enhancing botanical diversity within field margins of intensively managed grassland: a 7-year field experiment. J Appl Ecol 48(3):551–560

Garratt MP, Senapathi D, Coston DJ et al (2017) The benefits of hedgerows for pollinators and natural enemies depends on hedge quality and landscape context. Agr Ecosyst Environ 247:363–370

Hovd H, Skogen A (2005) Plant species in arable field margins and road verges of central Norway. Agr Ecosyst Environ 110(3–4):257–265

Marshall EJP (1987) Herbicide effects on the flora of arable field boundaries. In: Proceedings of the British crop protection conference-weeds, British Cop Protection Council, London, 16–19 November 1987

Marshall EJP (2004) Agricultural landscapes: field margin habitats and their interaction with crop production. J Crop Improv 12(1–2):365–404
Nautiyal S, Goswami M, Nidamanuri RR et al (2020) Structure and composition of field margin vegetation in the rural-urban interface of Bengaluru, India: a case study on an unexplored dimension of agroecosystems. Environ Monit Assess 192:520. https://doi.org/10.1007/s10661-020-08428-6
Pandey DN (2002) Carbon sequestration in agroforestry systems. Clim Policy 2(4):367–377
Rands SA, Whitney HM (2011) Field margins, foraging distances and their impacts on nesting pollinator success. PLoS ONE 6(10):1–8
Smith J, Potts SG, Woodcock BA et al (2008) Can arable field margins be managed to enhance their biodiversity, conservation and functional value for soil macrofauna? J Appl Ecol 45(1):269–278
Zhu W, Wang S, Caldwell CD (2012) Pathways of assessing agroecosystem health and agroecosystem management. Acta Ecol Sin 32(1):9–17

Part VI
Conclusion

Chapter 20
Dynamics in the Rural–Urban Interface of an Indian Megacity—Synopsis and Outlook

Ellen Hoffmann, Stephan von Cramon-Taubadel, and Andreas Buerkert

Abstract This editorial concludes the volume by reviewing the key findings of the contributions and putting them in context to each other. It highlights the challenges and merits of interdisciplinary approaches and sketches an outlook for future research.

Keywords Rural–urban transformations · Land use changes · Agricultural production patterns · Intensification · (Food) consumption patterns · Biodiversity · Social–ecological systems

20.1 Introduction

The studies compiled in this volume address rural–urban transformation processes related to urbanisation from different disciplinary angles, but all of them target the South Indian metropolitan region of Bengaluru. This allows exploring the multiple interrelations between economic, agronomic, ecological and social subsystems. The methodological approaches span from household surveys, vegetation/biodiversity assessments and field experiments to remote sensing and GIS-based mapping.

E. Hoffmann (✉) · A. Buerkert
Organic Plant Production and Agroecosystems Research in the Tropics and Subtropics, University of Kassel, 37213 Witzenhausen, Germany
e-mail: ellen.hoffmann@uni-kassel.de

A. Buerkert
e-mail: buerkert@uni-kassel.de

S. von Cramon-Taubadel
Agricultural Policy, Agricultural Economics and Rural Development, Georg-August-University Göttingen, 37073 Göttingen, Germany
e-mail: scramon@gwdg.de

E. Hoffmann et al. (eds.), *The Rural-Urban Interface*, The Urban Book Series,
https://doi.org/10.1007/978-3-030-79972-4_20

20.2 Synopsis

20.2.1 Changing Patterns of Land Use and Agriculture

Growing cities encroach their surrounding hinterlands and displace previous land cover by settlements and urban infrastructures, whereby the spatial patterns of this process vary as they depend on the prevailing geographic and sociopolitical context (Bagan and Yamagata 2014; Elmqvist et al. 2013). The papers in Part II combine a remote sensing perspective of land cover and land use changes in the rural–urban interface of Bengaluru with socio-economic household surveys, aiming to distinguish characteristics typical for urban, transitional or rural areas. In summary, it could be shown that transformation processes tend to develop high dynamics in rather localized spaces and yet have enough impact to change average characteristics over larger surrounding areas. Nuclei of urbanisation may emerge even in villages far from the city, while the triggers and drivers that link this development to the mega-urban core remain to be elucidated. With increasing urbanisation, the land use patterns show decreases in natural vegetation, agricultural area and fallows. Cropping patterns are characterised by a loss in area under staple crops with corresponding increases in the cultivation of horticultural and commercial crops, cropping intensity and diversification. At the household level, farming systems shifted from non-vegetable to vegetable systems, and increasingly included livestock, whereby the availability of credit and capital enhanced these developments.

20.2.2 Resource Use, Efficiencies and Tradeoffs in Crop and Livestock Production

Agricultural intensification and diversification, which were evident in Part II at a larger analytical scale, play out on the ground in terms of rising yields, improved efficiencies or use of basic resources—up to their potential overuse and degradation (Foley et al. 2011; Steffen et al. 2015). Experimental approaches at the field scale are required to analyse the underlying biophysical and physiological processes. Exemplary for the agronomic sector, Part III presents a factorial field experiment designed to model nutrient and field water cycles at different management intensities under the local, agroclimatic conditions in Bengaluru. These experiments deliver high data density at the microscale, and as a long-term setup they are linked to the other approaches by receiving and providing context and calibration. The livestock sector was exemplified by an on-farm study on the productivity and health of dairy cattle. Various modelling approaches will help to understand the nutrient flows, potential unintended side effects, thresholds, feedback mechanisms and trade-offs related to the changes observed at the household and regional scale. The production focus positions this chapter at the supply side of (peri-)urban food systems.

20.2.3 Food Systems as Outcomes and Drivers of Rural-Urban Transitions

On the consumption side of urban food systems in India, contradictory trends towards undernutrition and obesity coexist, often referred to as the "double burden of malnutrition" (Deaton and Drèze 2009). Their interlinkage with the agricultural production patterns is analysed in Part IV by relating consumption and nutrition patterns to urbanisation as a common driver. As in the previous chapters, the positioning along the rural–urban gradient is substituted for a true time series to capture transition processes. In summary, the crop diversity documented in Parts I and II is mirrored in the diversity of the food basket in the research area. Overall, the sampled population around Bangaluru has a higher nutritional status than the Indian average. Our data show that the diversity of diets increases towards more urban areas. Nevertheless, the double burden of malnutrition is also evident around Bengaluru. The spatial prevalence of undernutrition and several specific nutrient deficiencies is higher in rural areas, coinciding with a lower socio-economic status of the households and primarily agricultural livelihoods. This is particularly obvious among women and children. Over-nutrition and its associated non-communicable diseases become more frequent towards the urban area, coinciding with a consumption pattern suggestive of a nutrition transition towards so-called Western diets as a consequence of urbanisation. Again, women are most affected, and the trend is stronger in higher social strata. At the same time, awareness is rising especially among middle-class women, to (re)turn to healthier food choices, which they often perceive to be met in the traditional Indian food culture. The demand for healthy products goes along with increasing interest of the consumers in the modes of food production. This will eventually feed back into the farm management and has some potential to counterbalance, at least partially, unsustainable agricultural intensification.

20.2.4 Biodiversity and Ecosystem Functions under Rural-Urban Transitions

Although most of the fertile land around Bengaluru is cultivated (to feed the city), food production is only one of several ecosystem services that agricultural landscapes deliver (MEA 2005). The papers in Part V assess the biodiversity in agro-ecosystems and analyse the functions and services that depend on it in different land cover and land use contexts. Supporting, regulating and cultural services may be challenged if biodiversity decreases as a trade-off for maximizing production output. All of the biodiversity assessments presented here report clear changes in community structures along the rural–urban gradient, but the profiles in species diversity and species richness vary for different groups of organisms. While trees, flowering plants and bees show the highest diversity and/or richness in the urban area, the transitional areas have the highest species richness of birds, and field margin vegetation is most

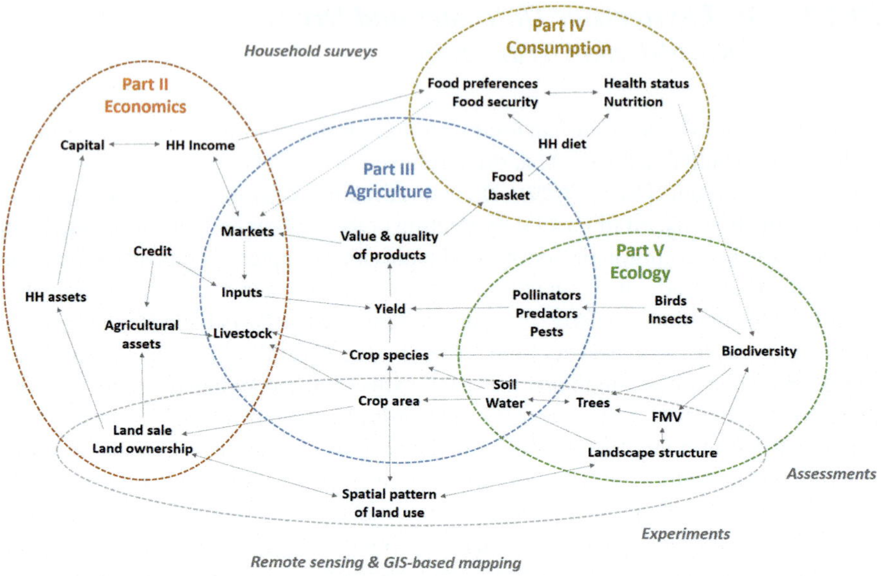

Fig. 20.1 Key issues of the papers in Parts II to IV and network of interrelations around agricultural transitions in the rural–urban interface of Bengaluru, India

diverse and most abundant in the rural area. Moreover, the local context, such as the extent of built-up land cover in the immediate surrounding of the assessment point, often has a stronger impact on the observed biodiversity than the large-scale position within the transect. Comprehensive biodiversity assessments thus provide site-specific data to validate, enrich and advance the current debate on sustainable urbanisation.

Agriculture is an arena in which economics, consumption and ecology intersect, and thus lends itself as an ideal starting point for the comprehensive analysis of social–ecological systems and their urbanisation-driven transitions (Fig. 20.1). While this network of interdependencies is certainly neither complete nor the only possible configuration, it illustrates the complexity of the scenario and the central significance of agricultural transitions.

20.3 Discussion and Outlook

Unravelling this complexity and understanding mutual dependencies of processes that are traditionally investigated in different scientific disciplines requires system thinking in a coherent, overarching framework. The contributions compiled in this volume exemplify this idea. In that sense, the Indo-German Research Unit FOR2432 has elaborated a conceptual and methodological toolkit that can be transferred to other hotspots of urbanisation in Asia and possibly beyond, even if those obviously have profoundly different site-specific contexts.

It has to be noted, however, that the interdisciplinary potential of the results presented here has not yet been fully harnessed. The results presented here provide a snapshot of key insights achieved after an initial three-year research period.

As the data from different methodological approaches become available at different speeds, their integration and synthesis are still going on. A long-term time horizon is thus essential for this kind of research. Within the ongoing second funding phase, progress will be made towards achieving this task.

This intermediate synopsis has revealed that there is indeed a transitional zone between the rural and the urban space with distinct and sometimes unique features, although it is hard to delineate clear boundaries. The diversity and dynamics of this interface often surpass that of the rural and urban areas, and thus, it holds high potential for stimulating development in different directions. This synopsis also allows us to identify additional topics and concepts that would further link the different sections and can strengthen integrated analysis of key resources and concepts such as water, One Health or polycentricity. Different modelling approaches are required to test and validate causal relations and feedback loops. Scenario development and simulations may thus become possible in the near future to support policies and planning towards sustainable urbanisation.

Acknowledgements The authors gratefully acknowledge the financial support provided by the German National Science Foundation, DFG, through grant number BU 1308/14-1, and by the Department of Biotechnology, DBT, India, through grant DBT/IN/German/DFG/14/BVCR/2016 as parts of the Indo-German consortium of DFG Research Unit FOR2432/1. We express our deep appreciation and thanks to all project leaders that collaborated in this project across different academic, administrative and cultural traditions. Without their open-minded attitude, patience and perseverance, this endeavour could not have been successful.

References

Bagan H, Yamagata Y (2014) Land-cover change analysis in 50 global cities by using a combination of Landsat data and analysis of grid cells. Environ Res Lett 9(6):1–13

Deaton A, Drèze J (2009) Food and nutrition in India: facts and interpretations. Econ Pol Wkly 44(7):42–65

Elmqvist T, Redman CL, Barthel S, Costanza R (2013) History of urbanization and the missing ecology. In: Emlqvist T et al (eds) Urbanization, biodiversity and ecosystem services: challenges and opportunities. Springer, Berlin

Foley JA, Ramankutty N, Brauman KA et al (2011) Solutions for a cultivated planet. Nature 478(7369):337–342

MEA—Millennium Ecosystem Assessment (2005) Millennium ecosystem assessment synthesis report. Island Press, Washington, DC

Steffen W, Richardson K, Rockström J et al (2015) Planetary boundaries: guiding human development on a changing planet. Science 347(6223):1259855

Printed by Printforce, the Netherlands